Feed-in Tariffs and the Economics of Renewable Energy

Yoshihiro Yamamoto

Feed-in Tariffs and the Economics of Renewable Energy

Springer

Yoshihiro Yamamoto
Takasaki City University of Economics
Takasaki
Japan

ISBN 978-3-030-08312-0 ISBN 978-3-319-76864-9 (eBook)
https://doi.org/10.1007/978-3-319-76864-9

Softcover re-print of the Hardcover 1st edition 2018

Printed on acid-free paper

This Springer imprint is published by the registered company Springer International Publishing AG part of Springer Nature
The registered company address is: Gewerbestrasse 11, 6330 Cham, Switzerland

Preface

Renewable energy sources, such as solar, wind, and biomass, are being developed worldwide. In addition to technological development, this is attributable to promotion by governments through policy instruments.

This book examines some economic and policy issues in the promotion of renewable energy. The first part of this book proposes an analytical model for investigating feed-in tariffs, a policy instrument for promoting renewable energy. It begins by reviewing several models, which are deduced from the models for investigating policy instruments that aim to reduce greenhouse gas emissions. However, they might not embody a critical aspect of feed-in tariffs: encouraging investment rather than increasing production in terms of electricity generated from renewable energy sources. Thus, the first part of the book presents alternative models.

In the second part, the book examines some important features of renewable energy development besides feed-in tariffs. They include uncertainty, engineering points of view, diffusion of innovation, partnership among relevant parties, and community. The second part offers different investigations into the promotion of renewable energy from economic and social perspectives.

This book takes a theoretical approach. It is possible to divide the study of promotion of renewable energy, including feed-in tariffs, into two categories: reports on the development of renewable energy and policies in various countries, and numerical investigations, including regression analysis and simulation. In contrast, few books approach these issues theoretically, particularly from an economic point of view. This book seeks to contribute theoretical investigations to a knowledge base.

This book is based on the research I have conducted thus far. I am grateful to numerous colleagues, conference participants, and students who have shaped my approach through comments and questions. I gratefully acknowledge the financial support by JSPS KAKENHI grants 24560500 and 17K00693.

Takasaki, Japan Yoshihiro Yamamoto

Contents

Chapter 1
Introduction

Abstract Renewable energy sources (RES) have been developed worldwide. Their rapid development can be attributed to, in addition to technological developments, various types of tools governments offer to support the use of RES, including the renewables portfolio standard (RPS) and feed-in tariffs (FITs). The book seeks to provide insights into such economic and policy issues. The purpose of this chapter is to present an overview of the book. Before briefly describing each chapter, we present Japanese renewable energy policy, which may provide a useful lesson for other countries because it has shifted from an RPS system to FITs. After reviewing it, we present an overview of this book. The book consists of two parts. Part I of the book conducts theoretical investigations into FITs by developing several simple microeconomic models. Part I consists of six chapters. Part II of the book addresses some economic and policy issues surrounding the development of RES. Part II consists of five chapters.

Keywords Feed-in tariff · Renewable energy promotion · Japanese renewable energy policy

1.1 The Use of Renewable Energy Sources

Renewable energy sources (RES) such as solar, wind, and biomass have been developed worldwide because they may mitigate global warming, alleviate energy security issues, enhance energy source diversity, create new business opportunities, and provide other benefits.

The rapid development of RES that is currently underway can be attributed to technological developments such as the generation of solar photovoltaic (PV) power and wind power. However, the installed capacity of renewable energy technologies is still smaller than what is needed, mainly because the cost of RES is still too high and thus impedes competition with conventional energy sources such as coal, oil, and gas.

Y. Yamamoto, *Feed-in Tariffs and the Economics of Renewable Energy*,
https://doi.org/10.1007/978-3-319-76864-9_1

Thus, governments offer various types of tools to support the use of RES, including tax credits, investment or capital subsidies, the renewables portfolio standard (RPS), and feed-in tariffs (FITs). Among these tools, FITs and RPS are set up specifically to promote the use of RES. In a typical RPS system, electricity retailers are forced to sell a set amount of RES-E. To fulfill their obligations, retailers may either generate RES-E on their own or purchase it from others. On the other hand, in a standard FIT system, a government sets a price at which a household or firm can sell the electricity generated from RES (RES-E) during a set period of years. Following the success of FITs in promoting the development of RES-E, FITs are attracting considerable attention from many governments.

However, few studies examine FITs theoretically; most existing studies are based on empirical investigations or discuss relevant policy issues by using survey reports. Successful development of FITs cannot be expected unless a FIT system is designed on theoretical grounds. Therefore, Part I of this book presents theoretical investigations into FITs by developing several simple microeconomic models.

It should be noted that the core of designing a FIT system is setting a price for RES-E to encourage investment in RES-E generation. However, it is not sufficient to examine the price of RES-E exclusively. Other factors are important as well: for example, a government may have to design a policy instrument under uncertainty in order to encourage investment by foreign firms; a government may account for a cost that is more likely to arise when a large quantity of RES-E is fed into the power grid; something other than the price of RES-E may play a role in the diffusion of RES-E generation; and, the organizational form and involvement of a community may matter to the development of RES. These issues surrounding the development of RES will be addressed in Part II of this book.

The purpose of this chapter is to present an overview of the book. Before briefly describing each chapter, we present Japanese renewable energy policy, which may provide a useful lesson for other countries because it has shifted from the quantity-based approach of RPS to the price-based approach of FITs.

The remainder of the chapter is organized as follows. In Sect. 1.2, we examine Japanese renewable energy policy. In Sect. 1.3, we present an overview of Part I of the book, which conducts theoretical investigations into FITs by developing several simple microeconomic models. In Sect. 1.4, we present an overview of Part II, which addresses some economic and policy issues surrounding the development of RES.

1.2 Renewable Energy Policy in Japan

Before presenting an overview of the book, it will be fruitful to examine the state of RES development in terms of policy implementation and technology diffusion. A large number of sources provide insight into RES development in various countries. In particular, renewable energy policy may be surveyed in books by Mendonça (2007), Mendonça et al. (2010), Ansuategi et al. (2015), Daim et al. (2015), Meier et al. (2015), and Mir-Artigues and del Río (2016). Many case studies have been conducted by Meier et al. (2015), including in Vietnam, Indonesia, South Africa, and Brazil in regard to incentive programs and economic and financial aspects of RES. Mir-Artigues and del Río (2016) also surveyed policy instruments used to promote PV generation in selected countries, such as the USA, Japan, Germany, Spain, the UK, and China. Furthermore, Mendonça (2007) and Mendonça et al. (2010) surveyed policy measures, focusing particularly on FITs and related programs for countries such as Germany, Spain, the UK, the USA, Canada, Australia, India, and South Africa.

Japanese renewable energy policy may provide a useful lesson for other countries because it has shifted from the quantity-based approach of RPS to the price-based approach of FITs. However, while information on the renewable energy policies of the USA, the UK, Germany, and Spain is relatively easy to access, information about Japanese renewable energy policy might be less available, especially to foreign researchers. Thus, it is worthwhile to survey Japan's policy here, although the contents of this book are not limited to the Japanese case.

Japanese renewable energy policy programs will be divided into two periods. During the first period, from April 2003 (and partially from December 2002) to June 2012, an RPS system was implemented. Electricity retailers, including 10 regional-monopolist utilities, were required to sell a target quantity of RES-E. The RES eligible for the RPS included solar, wind, geothermal, hydro, and biomass power generation. The government set the target amount of RES-E every four years; the target was provided as the total amount of RES-E and not specified for each type of RES. For example, for fiscal year 2003, the target amount of RES-E was 7320 GWh in total and was broken down for each retailer based on its electricity sales. There were three means by which retailers could meet their obligations: generating RES-E on their own, purchasing RES-E from others, or purchasing a certificate proving a certain amount of RES-E generation.

The RPS terminated in June 2012, and a system of FITs has been in place ever since. One of the reasons for this change might be that while renewable energy technologies had been diffused, to some extent, because of the RPS in Japan, remarkable success was achieved with FITs in other countries such as Germany and Spain. The ongoing FIT allows households and businesses to sell their generated RES-E to an electric utility at a set price during a set number of years. As with the RPS, solar, wind, geothermal, hydro, and biomass power generation can all be applied to FITs. The prices are set for particular types of RES. For example, for PV generation, in 2017, a price of ¥28 or ¥30 is set for installed capacity below 10 kW for a period of 10 years; ¥21 is the set price for installed capacity of 10–2000 kW for a period of 20 years, and the price is put out to tender for installed capacity above 2000 kW (at present, $1.00 is approximately equal to ¥110). For wind-power generation on land, a price of ¥28 is set for installed capacity of no less than 20 kW, and a price of ¥55 is set for capacity below 20 kW.

It is noted that electric utilities had already purchased RES-E before the current FIT system commenced in 2012. From 1992 to 2009, they purchased voluntarily surplus RES-E, particularly surplus PV electricity, which is generated but not self-consumed by customer-generators, at a price approximately equal to the retail

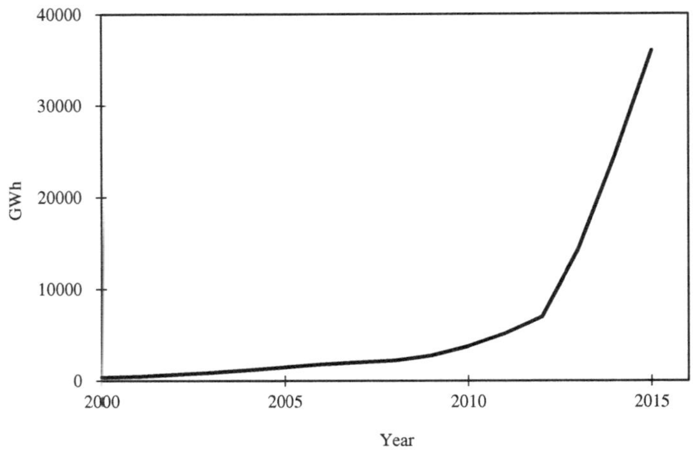

Fig. 1.1 Accumulation of the amount of PV generation. The amount of PV electricity has been rapidly increasing since 2012, when the current FITs began in place of the RPS. *Data* IRENA (2017)

electricity rate. Subsequently, from 2009 to 2012, the utilities were forced to purchase surplus PV electricity at a set price for a period of 10 years. Thus, the FITs for PV electricity coexisted with the RPS system from 2009 to 2012. Finally, the current FITs were implemented in place of the RPS in July 2012. These experiences might affect, to some extent, the transition from RPS to FITs in Japan.

Owing to these policy measures and to technological developments, the amount of RES-E in Japan has increased. Figures 1.1 and 1.2 show the accumulation of PV generation and wind-power generation, respectively. The amount of RES-E generation has been gradually increasing since approximately 2003, when the RPS was implemented. In particular, the amount of PV generation has been rapidly increasing since 2012, when the current FITs were implemented in place of the RPS.

1.3 Analysis of a Feed-in Tariff System

In this section, we present an overview of Part I of the book, which conducts theoretical investigations into FITs by developing several simple microeconomic models. Part I consists of six chapters.

In Chap. 2, FITs and RPS are comparatively modeled based on their similarities to policy instruments for reducing greenhouse gas emissions. On the one hand, FITs are a price-based policy tool in the sense that a government sets a price at which RES-E can be sold. On the other hand, RPS is a quantity-based policy tool in the sense that a government forces electricity retailers to sell a set amount of RES-E. Recall that for reducing greenhouse gas emissions, a carbon tax is a price-based tool, whereas tradable emission permits are a quantity-based tool. In this regard, it is often

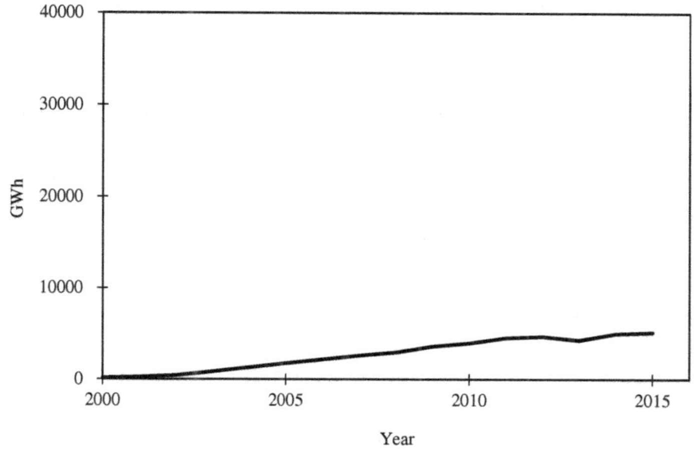

Fig. 1.2 Accumulation of the amount of wind-power generation. The amount of wind-power electricity has been gradually increasing since approximately 2003, when the RPS was implemented. *Data* IRENA (2017)

argued that FITs and RPS are comparable to carbon taxes and tradable emission permits, respectively. Accordingly, they may be modeled similarly to carbon taxes and tradable emission permits. Based on this conjecture, FITs and RPS are modeled in terms of marginal conditions, optimization, and linear programming, one by one, in Chap. 2.

In Chap. 3, we develop a microeconomic model for investigating FITs, which may be different from the model in Chap. 2 that is deduced from the model for investigating a carbon tax. We should note two features of generating RES-E that are distinct from the features of reducing greenhouse gases. First, the amount of RES-E output cannot be controlled; it depends, to a large extent, on natural conditions. Second, fixed investment costs are much more important than variable operating costs. Hence, we need to develop a new type of model to investigate FITs. We consider PV generation in the business sector and in the residential sector. The model pays particular attention to heterogeneity among decision-makers: given a price of PV electricity under FITs, some decision-makers will invest in PV generation and others will not. Using this model, we examine the amount of PV electricity generated and address the problem of social welfare maximization.

In Chap. 4, we compare three types of FITs. To incentivize households to adopt a PV system, there are three types of FITs, each of which prices a different part of PV electricity: all PV electricity, surplus PV electricity, and the difference between PV generation and electricity consumption. In this chapter, we refer to these as FITs for all PV electricity, FITs for surplus PV electricity, and net metering, respectively. The study aims to compare these mechanisms with respect to retail electricity rates, including the cost to an electric utility of purchasing PV electricity, and with respect to social welfare. A microeconomic model is developed, and the results are confirmed by means of a simulation. If we account for some reductions in electricity consumption with FITs for surplus PV electricity or net metering, the results for social welfare should be slightly modified. Chapter 4 is based on the study by Yamamoto (2012).

In Chap. 5, a combination of FITs and capital subsidies is investigated. Both FITs and capital subsidies have been widely employed to promote the adoption of renewable energy technologies, and this chapter sheds light on the combined use of both tools. The purpose is to clarify how these tools can be optimally combined to encourage households to adopt PV systems or firms to invest in PV generation. The study develops a microeconomic model embodying the idea of two-part tariffs. Maximization of PV electricity to be generated, minimization of promotion cost, and maximization of social welfare are examined. In particular, the FIT level that maximizes social welfare is identified. Most of Chap. 5 draws on the study by Yamamoto (2017).

In Chap. 6, we conduct simulations to confirm the results and obtain new insights into findings that were unclear in Chap. 5, which studies optimal combinations of FITs and capital subsidies. We conduct simulations for the adoption of PV systems in the residential sector. Parameter values are set based on a variety of data sources. The simulations verify the theoretical results of Chap. 5 and provide new findings with regard to the amount of PV electricity, the promotion cost, and social welfare. The comparison between the two cases, FITs for all PV electricity and FITs for surplus PV electricity, provide some useful results. Part of Chap. 6 is based on the study by Yamamoto (2017).

In Chap. 7, a variant of the model in Chap. 5 is presented. In Chap. 5, we have developed a microeconomic model to investigate optimal combinations of FITs and capital subsidies for the adoption of PV systems in the residential sector. In that model, it was assumed that a household, a potential adopter, is characterized by several variables related to PV generation. The variables were discrete so that an individual household could be examined with regard to the adoption of a PV system. In contrast, the variables are described as continuous in Chap. 7. In this model, a government controls FIT and capital subsidy levels to attain a target quantity of adoption. By using this model, we consider three optimality criteria: maximization of PV electricity, minimization of promotion cost, and maximization of social welfare. The same results are obtained as in Chap. 5 with respect to the optimal combinations of FITs and capital subsidies.

1.4 Economic and Policy Issues of Renewable Energy

Part II of the book addresses some issues surrounding the development of RES. In this section, we present an overview of Part II, which consists of five chapters.

In Chap. 8, we consider uncertainty in modeling a combination of feed-in premiums (FIPs) and capital subsidies. Foreign direct investment in renewable energy projects, in particular where biomass is used as input, has been attracting increasing attention. In the case of foreign direct investment, there may be an information gap between a host country's government and the foreign firm that will invest: while the firm can collect information regarding the project through a feasibility study, for example, it will be difficult for the government to know whether a foreign firm is undertaking the project efficiently. It is assumed that the government will offer the foreign firm some remunerations—consisting of FIPs and capital subsidies—to encourage investment in such a project. The purpose of this chapter is to determine the optimal combination of FIPs and capital subsidies that encourages investment in a renewable energy project by a foreign firm. To this end, we develop a microeconomic model that accounts for this information gap. The model developed in this chapter may be considered to extend the model developed in Chap. 5 to an investigation that takes uncertainty into account.

In Chap. 9, a new perspective is offered on the modeling of pricing the RES-E that is fed into the power grid. As an increasing amount of RES-E is fed into the grid, various problems occur more frequently, such as frequency and voltage instability. To address this problem, a system operator provides ancillary services such as balancing supply and demand for electricity and procuring reactive power supply. Then, the cost of ancillary services should be appropriately allocated to distributed generators of RES-E. This chapter proposes a method for solving this cost allocation problem. The method proposed is an application of the Aumann–Shapley rule, which is one of cost-sharing rules among multiple entities. The method may be useful for designing a new type of feed-in tariff system, which will be needed after a diffusion goal is achieved under the current FIT system.

In Chap. 10, the adoption of PV systems in society is examined by means of diffusion theory. According to diffusion theory, opinion leaders play an important role in the diffusion of new technologies through interpersonal communication with potential adopters. The purpose of this chapter is to examine whether this is the case for a PV system and to investigate the role and utility of opinion leadership in its diffusion. The study employed an internet-based questionnaire survey to assess the use of interpersonal communication in decision-making on adoption, to identify opinion leaders with respect to adoption and to characterize their WTP. The response pool consisted of 488 individuals who lived in detached houses in Japan, owned a residential PV system and were responsible for making the decision to adopt their PV system. Chapter 10 draws on the study by Yamamoto (2015).

In Chap. 11, we are concerned with public-private partnership (PPP) in a renewable energy project. An increasing number of projects in which biomass discarded as waste, called biomass waste in this chapter, is utilized as a renewable resource have been implemented worldwide. This type of project, called a biomass project, will involve various types of parties, such as municipalities, private companies, consortia, and NGOs. The purpose of this chapter is to clarify the optimal organizational form of a biomass project. To begin, we survey cases of biomass projects in Japan to identify their typical features. Considering that a biomass project has both public and commercial aspects, we are concerned with PPP as an organizational form for biomass projects. To examine the applicability of PPP to a biomass project, we review previous studies of PPP in the economics literature.

In Chap. 12, the role of a community in developing RES is investigated. Many renewable energy (RE) cooperatives, particularly in Europe, undertake local RE projects such as PV generation and wind-power generation. In contrast, in Japan, a municipality has recently become actively involved in setting up a company that is undertaking such a project. This type of company will be called "a municipal RE company" in this chapter. The purpose of the chapter is to examine the effectiveness of the organizational form of a municipal RE company. A literature review, website surveys and an interview are conducted. It is suggested that a municipal RE company works, to some extent, in the same way as an RE cooperative and thus may be effective at undertaking local RE projects. Most of Chap. 12 draws on the study by Yamamoto (2018).

References

Ansuategi A, Delgado J, Galarraga I (eds) (2015) Green energy and efficiency: an economic perspective. Springer International Publishing, Cham

Daim TU, Kim J, Iskin I, Abu Taha A, van Blommestein KC (eds) (2015) Policies and programs for sustainable energy innovations: renewable energy and energy efficiency. Springer International Publishing, Cham

IRENA (2017) Data and statistics. resourceirena.irena.org/gateway/dashboard/. Accessed 30 Sept 2017

Meier P, Vagliasindi M, Imran M, Eberhard A, Siyambalapitiya T (2015) The design and sustainability of renewable energy incentives: an economic analysis. International Bank for Reconstruction and Development/The World Bank, Washington

Mendonça M (2007) Feed-in tariffs: accelerating the deployment of renewable energy. Earthscan, London

Mendonça M, Jacobs D, Sovacool B (2010) Powering the green economy: the feed-in tariff handbook. Earthscan, New York

Mir-Artigues P, del Río P (2016) The economics and policy of solar photovoltaic generation. Springer International Publishing, Cham

Yamamoto Y (2012) Pricing electricity from residential photovoltaic systems: a comparison of feed-in tariffs, net metering, and net purchase and sale. Sol. Energy 86:2678–2685

Yamamoto Y (2015) Opinion leadership and willingness to pay for residential photovoltaic systems. Energy Policy 83:185–192

Yamamoto Y (2017) Feed-in tariffs combined with capital subsidies for promoting the adoption of residential photovoltaic systems. Energy Policy 111:312–320

Yamamoto Y (2018) Optimal organizational forms for local renewable energy projects. In: Sayigh A (ed) Transition towards 100% renewable energy: selected papers from the World Renewable Energy Congress WREC 2017, Chap. 42, Springer International Publishing, Cham

Part I
Analysis of a Feed-in Tariff System

Chapter 2
Feed-in Tariffs in Comparison with the Renewables Portfolio Standard

Abstract Feed-in tariffs (FITs) and the renewables portfolio standard (RPS) are two major policy instruments for promoting the development of the electricity generated from renewable energy sources (RES-E). On the one hand, FITs are a price-based policy tool in that a government sets a price at which RES-E can be sold for a set period of years. On the other hand, RPS is a quantity-based policy tool in that a government forces electricity retailers to sell a set amount of RES-E. Recall that a carbon tax is a price-based tool, whereas tradable emission permits are a quantity-based tool for reducing greenhouse gas emissions. In this regard, it is often argued that FITs and RPS are comparable to carbon taxes and tradable emission permits, respectively. Accordingly, they may be modeled similarly to carbon taxes and tradable emission permits. Based on this conjecture, FITs and RPS are modeled in this chapter in terms of marginal conditions, optimization, and linear programming, one by one. However, we should note two features of generating RES-E that are distinct from reducing greenhouse gases. First, the amount of RES-E output cannot be controlled; it depends, to a large extent, on natural conditions. Second, fixed investment costs are much more important than variable operating costs. Hence, we need to develop an alternative model to investigate FITs and RPS.

Keywords Feed-in tariff · Renewables portfolio standard · Carbon tax
Tradable emission permit · Optimization

2.1 Introduction

There are two types of systems, feed-in tariffs (FITs) and the renewables portfolio standard (RPS), that have been widely used by governments to promote the development of electricity generated from renewable energy sources (RES-E).

In a standard FIT system, a government offers a price at which an adopter of renewable energy technology can sell the electricity generated during a set period of years. An electric utility or electricity distribution company must purchase that electricity, while it can add the cost of purchasing on the retail electricity rate. On

Y. Yamamoto, *Feed-in Tariffs and the Economics of Renewable Energy*,
https://doi.org/10.1007/978-3-319-76864-9_2

the other hand, in a standard RPS system, electricity retailers are obliged to sell a set amount of RES-E. In complying with that obligation, they can trade obligations; electricity retailers are allowed to purchase RES-E or certificates of RES-E generation from others, in addition to generating RES-E on their own. Hence, we may summarize as follows: in a FIT system, a government sets the price of RES-E to be purchased, whereas in an RPS system, it sets a quantity of RES-E to be supplied.

These characteristics of FITs and RPS may be comparable to two primary policy measures for reducing greenhouse gases, that is, carbon taxes and tradable emission permits (Menanteau et al. 2003). As is well known, in a carbon tax system, a firm must pay tax according to its amount of carbon emissions. On the other hand, in a tradable emission permits system, a firm must have a quantity of permits that matches the quantity of its carbon emissions; the firm reduces carbon emissions on its own or purchases emission permits from others. Thus, in summary, a price is set in a carbon tax system, whereas a quantity is set in a tradable emission permits system (Mankiw 2011). Hence, it may be stated that FITs are comparable to carbon taxes, while RPS is comparable to tradable emission permits. Accordingly, FITs and RPS may be understood analogously to carbon taxes and tradable emission permits. Furthermore, considering that—as is well known—carbon taxes and tradable emission permits are identical in theory, it may be the case that FITs and RPS are theoretically identical.

The purpose of this chapter is to clarify the mechanisms of FITs and RPS, thereby showing that the two systems are identical in theory. To this end, we develop different types of models, which are similar to the models of carbon taxes and tradable emission permits. For analytical simplicity, it is assumed in our models in the following sections that in an RPS system, firms will comply with their RES-E supply obligations by generating RES-E on their own or by purchasing RES-E on the RES-E wholesale market; we do not consider the case where firms purchase RES-E directly from others and where they purchase certificates of RES-E generation.

The remainder of the chapter is organized as follows. In Sect. 2.2, we present a model of FITs and RPS in terms of marginal conditions. This modeling is simple but does not necessarily reflect the mechanisms of, in particular, RPS. In Sect. 2.3, we then provide an alternative model, which is in terms of optimization. The similarity between the two systems will be clearer in a model presented in terms of linear programming, as discussed in Sect. 2.4. Section 2.5 concludes the chapter.

2.2 Modeling in Terms of Marginal Conditions

It may be possible to model FITs and RPS by means of marginal conditions, with which carbon taxes and tradable emission permits are studied. According to Menanteau et al. (2003), in a FIT system, producers are encouraged to exploit available sites for RES-E generation, e.g., wind energy, until their marginal cost equals the offered FIT level. On the other hand, in an RPS system, every producer exploits available sites until its marginal cost equals the equilibrium price of the certificate and then either purchases certificates to attain the assigned amount of RES-E or sells surplus certificates on the market. Menanteau et al. explained these mechanisms with graphical illustrations.

As Menanteau et al. (2003) argue, this way of understanding FITs and RPS is similar to that of understanding carbon taxes and tradable emission permits. Thus, let us develop a model for FITs and RPS that is similar to the model for the carbon tax and tradable emission permits. Following the reasoning by Menanteau et al. (2003), we develop a model in terms of marginal conditions; it is based on the model by Kolstad (2000) of emission fees and marketable ambient permits for emissions reduction.

Let q be the amount of RES-E a firm supplies.

Define the cost function of the firm as $C(q)$. It is assumed that $C(q)$ is a C^2 function on $\mathbf{R}^1_+$ such that $C'(q) > 0,\ C''(q) > 0$.

2.2.1 The Model of FITs in Terms of Marginal Conditions

First, let us examine a FIT system. Suppose that a government sets a FIT level at p: a firm can sell RES-E to an electric utility at p. The firm's profit maximization problem is

$$\text{maximize} \;\; pq - C(q). \tag{2.1}$$

If we assume the existence of inner solutions, the first-order condition yields $MC(q^*) = p$, where $MC(q) \equiv C'(q)$ is the firm's marginal cost function. In other words, it should supply an amount of RES-E so that the marginal cost equals the FIT level.

2.2.2 The Model of RPS in Terms of Marginal Conditions

Next, let us examine an RPS system. The government allocates to a firm obligations $\bar{q}$ of supplying RES-E at the outset. Then, the firm would minimize the cost of fulfilling the obligation. If it is assumed that the RES-E wholesale market is competitive, the firm's cost minimization problem is

$$\text{minimize} \;\; C(q) + \rho \cdot (\bar{q} - q), \tag{2.2}$$

where ρ is the equilibrium price on the RES-E wholesale market. The firm pays $\rho(\bar{q} - q)$ if it generates q less than $\bar{q}$ and purchases $\bar{q} - q$ on the market; it gains $\rho(q - \bar{q})$ otherwise if it generates q more than $\bar{q}$ and sells the surplus $q - \bar{q}$ on the market. In both cases, the cost to the firm is $C(q) + \rho \cdot (\bar{q} - q)$.

The first-order condition yields $MC(q^*) = \rho$. The firm produces q^* on its own and either purchases $\bar{q} - q^*$ on the market if $q^* < \bar{q}$ or sells $q^* - \bar{q}$ if $q^* > \bar{q}$. In other words, in this model, a firm should determine the amount of RES-E supply by equating the marginal cost of supplying RES-E with the equilibrium price on the RES-E wholesale market. This is essentially the same condition as the condition for a FIT system.

It is argued from this investigation that setting a FIT level at p, so that the total supply of RES-E equals $\bar{q}$, is equivalent to setting an RPS obligation at $\bar{q}$ so that the equilibrium price of the RES-E equals p on the wholesale market (Menanteau et al. 2003).

It should be noticed that the above way of modeling does not directly describe a firm's decision-making under an RPS system. Recall that in our assumption, RPS forces a firm to supply a certain amount of RES-E, either by producing RES-E on its own or procuring RES-E from an RES-E wholesale market. The firm, on the other hand, fulfills its obligations while minimizing the cost of doing so. In contrast, in the above modeling, given an equilibrium price on the RES-E wholesale market, the firm determines how much of RES-E to produce on its own, and how much to purchase on the market, by minimizing the cost of fulfilling the obliged amount of RES-E. There is a subtle difference between the two. Next, more direct modeling will be presented in terms of optimization.

2.3 Modeling in Terms of Optimization

In this section, we develop a model of FITs and RPS in terms of optimization. In particular, RPS is modeled more directly, reflecting a firm's decision-making. This way of modeling will make the relationship between FITs and RPS much clearer.

Let $\mathbf{x}$ denote an input vector of order n, each element x_i of which represents inputs for RES-E generation, such as land, equipment, and labor.

Let w_i be a unit cost of input x_i. Hence, $\mathbf{w} \cdot \mathbf{x}$ is the firm's cost of supplying RES-E.

Define $f(\mathbf{x})$ as the amount of RES-E the firm generates using input $\mathbf{x}$. It is assumed that $f(\mathbf{x})$ is a C^2 function on $\mathbf{R}^n_+$.

Generally, some constraints should be placed on input $\mathbf{x}$, reflecting the technology of RES-E generation and availability of inputs. However, such constraints—except for non-negativity constraints—are left out of the model for the sake of simplicity.

2.3.1 *The Model of FITs in Terms of Optimization*

First, suppose that a government adopts a FIT system, setting a price p at which a firm can sell the RES-E it generates. The firm's profit maximization problem is

$$\text{maximize} \quad pf(\mathbf{x}) - \mathbf{w} \cdot \mathbf{x}, \tag{2.3}$$

$$\text{subject to} \quad \mathbf{x} \geq \mathbf{0}. \tag{2.4}$$

The Lagrangian for this problem is

$$L(x_1, \ldots, x_n; \xi_1, \ldots \xi_n) = pf(x_1, \ldots, x_n) - \sum\nolimits_{i=1}^{n} w_i x_i + \sum\nolimits_{i=1}^{n} \xi_i x_i. \tag{2.5}$$

where ξ_i $(i = 1, \ldots, n)$ is a Lagrange multiplier.

The first-order conditions are

$$\frac{\partial L}{\partial x_i} = p\frac{\partial f}{\partial x_i}(\mathbf{x}) - w_i + \xi_i = 0,\ i = 1, \ldots, n; \tag{2.6}$$

$$\xi_i x_i = 0,\ i = 1, \ldots, n; \tag{2.7}$$

$$\text{and } \xi_1 \geq 0, \ldots, \xi_n \geq 0, \tag{2.8}$$

along with the original constraints (2.4).

From (2.6), for every i,

$$p\frac{\partial f}{\partial x_i}(\mathbf{x}^*) - w_i^* = -\xi_i^* \leq 0. \tag{2.9}$$

Hence, $p\partial f/\partial x_i(\mathbf{x}^*) \leq w_i^*$ with $p\partial f/\partial x_i(\mathbf{x}^*) = w_i^*$ for all i such that $\xi_i^* = 0$ i.e., $x_i^* > 0$ from (2.7). This implies that a firm should equate the value of the marginal product with the input price (Varian 2014, p. 505).

2.3.2 *The Model of RPS in Terms of Optimization*

Next, suppose that the government adopts an RPS system, where a firm is allocated obligations to supply $\bar{q}$ of RES-E. Then, the firm must fulfill those obligations either by generating RES-E on its own or trading RES-E on the RES-E wholesale market. The firm's cost minimization problem is

$$\text{minimize } \mathbf{w} \cdot \mathbf{x}, \tag{2.10}$$

$$\text{subject to } f(\mathbf{x}) \geq \bar{q}, \tag{2.11}$$

$$\text{and } \mathbf{x} \geq \mathbf{0}. \tag{2.12}$$

The Lagrangian for this problem is

$$\tilde{L}(x_1, \ldots, x_n; \lambda; \zeta_1, \ldots, \zeta_n) = \sum_{i=1}^{n} w_i x_i - \lambda(f(x_1, \ldots, x_n) - \bar{q}) - \sum_{i=1}^{n} \zeta_i x_i. \tag{2.13}$$

where λ and ζ_i $(n = 1, \ldots, n)$ are Lagrange multipliers.

The first-order conditions are

$$\frac{\partial \tilde{L}}{\partial x_i} = w_i - \lambda \frac{\partial f}{\partial x_i}(\mathbf{x}) - \zeta_i = 0, i = 1, \ldots, n; \tag{2.14}$$

$$\lambda (f(\mathbf{x}) - \bar{q}) = 0; \tag{2.15}$$

$$\zeta_i x_i = 0, \quad i = 1, \ldots, n; \tag{2.16}$$

$$\lambda \geq 0; \tag{2.17}$$

$$\text{and} \ \ \zeta_1 \geq 0, \ldots, \zeta_n \geq 0, \tag{2.18}$$

plus the inequalities in (2.11) and (2.12).

From (2.14), we obtain

$$\lambda \frac{\partial f}{\partial x_i}(\mathbf{x}^*) - w_i = -\zeta_i \leq 0. \tag{2.19}$$

Hence, $\lambda\, \partial f/\partial x_i(x^*) \leq w_i$ with $\lambda\, \partial f/\partial x_i(x^*) = w_i$ for all i such that $\zeta_i = 0$, i.e., $x_i^* > 0$ from (2.16).

Considering $\bar{q}$ as a parameter, let $\mathbf{x}^*(\bar{q})$ denote the solution to the problem presented in (2.10) through (2.12), and let $\lambda^*(\bar{q})$ be the corresponding Lagrange multiplier. Then,

$$\lambda^*(\bar{q}) = \frac{d}{d\bar{q}} \mathbf{w} \cdot \mathbf{x}^*(\bar{q}) \tag{2.20}$$

holds (for example, Simon and Blume 1994, p. 451). This means that the multiplier $\lambda^*(\bar{q})$ represents the change in the minimized cost resulting from the unit change in the RES-E supply obligations. In other words, if the obligations are infinitesimally reduced, the minimized cost decreases by $\lambda^*(\bar{q})$; if the obligations are infinitesimally augmented, the minimized cost increases by $\lambda^*(\bar{q})$. Hence, the firm will determine the amount of RES-E it trades on the RES-E wholesale market—so that $\lambda^*(\bar{q})$ equals the equilibrium price of RES-E on the market—by either purchasing or selling RES-E on the market in order to change the RES-E amount it must produce on its own. Therefore, from $\lambda\, \partial f/\partial x_i(x^*) = w_i$ and Eq. (2.20), a firm should again equate the value of the marginal product with the input price.

Notice in the above calculations that the first-order conditions of the profit maximization problem in a FIT system are the same as those of the cost minimization problem in an RPS system. In this regard, the relationship between the two systems will be much clearer if we model FITs and RPS by means of linear programming in the next section.

2.4 Modeling in Terms of Linear Programming

Modeling in terms of linear programming will clarify the similarity between FITs and RPS more directly. The modeling utilizes the duality of the problems. Yamaji

(2007) used this idea to clarify the relationship between carbon taxes and tradable emission permits.

As above, $\mathbf{x}$ denotes a column vector of order n, each element x_i of which represents inputs for RES-E generation, such as land, equipment, and labor.

Let $\mathbf{a}$ be a row vector of order n, which defines the technology of RES-E generation in that $\mathbf{ax}$ is the output of RES-E when $\mathbf{x}$ is input.

Let $\mathbf{B}$ be an $m \times n$ matrix of the coefficients of constraints on the inputs and $\mathbf{b}$ be a vector of order m that specifies the availability of the inputs, that is, the bounds associated with the input constraints.

We reset cost vector $\mathbf{w}$, which was a column vector of order n in Sect. 2.3, to a row vector of order n in this section.

2.4.1 The Model of FITs in Terms of Linear Programming

Suppose that a government adopts a FIT system: a firm can sell RES-E at p. Then, the firm's profit maximization problem, which is designated the primal problem, is

$$\text{maximize} \quad p\mathbf{ax} - \mathbf{wx}, \tag{2.21}$$

$$\text{subject to} \quad \mathbf{Bx} \leq \mathbf{b}, \tag{2.22}$$

$$\text{and} \quad \mathbf{x} \geq \mathbf{0}. \tag{2.23}$$

The dual problem is

$$\text{minimize} \quad \boldsymbol{\omega}\mathbf{b}, \tag{2.24}$$

$$\text{subject to} \quad \boldsymbol{\omega}\mathbf{b} \geq p\mathbf{a} - \mathbf{w}, \tag{2.25}$$

$$\text{and} \quad \boldsymbol{\omega} \geq \mathbf{0}, \tag{2.26}$$

where$\boldsymbol{\omega}$is a row vector of order m.

The meaning of the dual problem is that if a buyer were to offer to the firm the purchase of the input resources, the buyer, seeking to minimize the cost of purchasing, would have to set prices $\boldsymbol{\omega}$ so that the gain of the firm by selling the resources should not be less than the loss of profit.

2.4.2 The Model of RPS in Terms of Linear Programming

Next, suppose that the government adopts RPS rather than FITs: a firm is forced to supply $\bar{q}$ of RES-E. The firm's cost minimization problem, which is designated the primal problem, is

$$\text{minimize} \quad \mathbf{wx}, \tag{2.27}$$

$$\text{subject to} \quad \mathbf{Bx} \leq \mathbf{b}, \tag{2.28}$$

$$\mathbf{ax} \geq \bar{q}, \tag{2.29}$$

$$\text{and} \quad \mathbf{x} \geq \mathbf{0}. \tag{2.30}$$

This problem can be transformed into the standard form of a maximization problem to

$$\text{maximize} \quad -\mathbf{wx}, \tag{2.31}$$

$$\text{subject to} \quad \mathbf{Bx} \leq \mathbf{b}, \tag{2.32}$$

$$-\mathbf{ax} \leq -\bar{q}, \tag{2.33}$$

$$\text{and} \quad \mathbf{x} \geq \mathbf{0}. \tag{2.34}$$

Then, the dual problem is

$$\text{minimize} \quad \boldsymbol{\pi}\mathbf{b} - \pi_a\bar{q}, \tag{2.35}$$

$$\text{subject to} \quad \boldsymbol{\pi}\mathbf{B} - \pi_a\mathbf{a} \geq -\mathbf{w}, \tag{2.36}$$

$$\boldsymbol{\pi} \geq \mathbf{0}, \tag{2.37}$$

$$\text{and} \quad \pi_a \geq 0. \tag{2.38}$$

where $\boldsymbol{\pi}$ is a row vector of order m and π_a is a scalar. Together, these constitute the dual variables.

Because $\pi_a\bar{q}$ is a scalar constant, the solution remains the same if $\pi_a\bar{q}$ is subtracted from the objective function (2.35). Then, the dual problem is expressed as

$$\text{minimize} \quad \boldsymbol{\pi}\mathbf{b}, \tag{2.39}$$

$$\text{subject to} \quad \boldsymbol{\pi}\mathbf{B} \geq \pi_a\mathbf{a} - \mathbf{w}, \tag{2.40}$$

$$\text{and} \quad \boldsymbol{\pi} \geq \mathbf{0}. \tag{2.41}$$

Looking at the dual problems of FITs, Eqs. (2.24–2.26), and RPS, Eqs. (2.39–2.41), we can see that the dual problems are exactly the same as each other. Because the dual of the dual problem is generally equivalent to the primal problem (for example, Ignizio and Cavalier 1994), the primal problems of FIT and RPS are identical.

2.5 Conclusion

Two policy tools, FITs and RPS, both of which have been widely used to promote the development of RES-E in society, were investigated to show the similarity between them. We developed three types of models—analogous to the models of environmental policy tools—because FITs and RPS are considered to have the same mechanisms as carbon taxes and tradable emission permits, respectively. The three types of modeling are in terms of marginal conditions, optimization, and linear programming. Every model showed that FITs and RPS are identical in theory.

The ways of modeling, however, may be criticized for not reflecting the reality of investments in projects such as wind-power and solar photovoltaic generation.

The most critical point is that it will be impossible for a firm to control the expected amount of RES-E output in a project. Generally, the capacity of electricity generation a firm sets up at a site is primarily determined by the availability of the land area. Furthermore, the amount of RES-E generated per unit capacity at a site during a certain period is determined by natural conditions such as wind intensity and solar radiation, none of which we can control. All we can do is to decide whether to invest in a site; we do not have any control over the amount of RES-E output after the investment.

In addition, it should be noted that the marginal cost of producing RES-E will be so small that we can consider it zero, especially for wind-power and photovoltaic generation.

Accordingly, the fixed investment cost is much more important than the variable operating cost. In other words, it is not appropriate to model FITs and RPS in the same way as carbon taxes and tradable emission permits, although it is often argued that FITs and RPS are very similar to carbon taxes and tradable emission permits, respectively. We need to develop a new type of modeling to examine FITs and RPS. Thus, in the following chapter, we develop a more appropriate model to study FITs.

References

Ignizio JP, Cavalier TM (1994) Linear programming. Prentice Hall, Upper Saddle River
Kolstad CD (2000) Environmental economics. Oxford University Press, Oxford
Mankiw NG (2011) Essentials of economics, 6th edn. South-Western, Mason
Menanteau P, Finon D, Lamy M-L (2003) Prices versus quantities: choosing policies for promoting the development of renewable energy. Energy Policy 31:799–812
Simon CP, Blume L (1994) Mathematics for economists. W.W. Norton & Company, New York
Varian H (2014) Intermediate microeconomics, 9th edn. W.W. Norton & Company, New York
Yamaji K (2007) Shisutemu suuri kougaku. Suurikougaku-sha, Tokyo (In Japanese)

Chapter 3
Modeling of Feed-in Tariffs

Abstract The similarity between feed-in tariffs (FITs) and carbon taxes is noted because both are price-based policy instruments, which makes it appropriate to model FITs similarly to carbon taxes when investigating renewable energy policy. However, generating electricity from renewable energy sources (RES-E) is fundamentally different from reducing greenhouse gas emissions, except for the difference between incentives and disincentives. Typically, whether to invest in renewable energy is much more relevant to FITs than is the quantity of RES-E to generate. Accordingly, we need to develop a model to investigate FITs, which may be different from the model used to investigate carbon taxes. The purpose of this chapter is to develop such a model. We consider solar photovoltaic (PV) power generation in the business sector and in the residential sector. The model pays particular attention to the heterogeneity among decision-makers: given a price of PV electricity under FITs, some decision-makers will invest in PV generation and others will not. Using the model, we examine the amount of PV electricity generated and address the problem of social welfare maximization. It is shown that while the amount of PV electricity increases as the price of PV electricity is set higher, there is a definite price at which social welfare is maximized.

Keywords Feed-in tariff · Business sector · Residential sector
Social welfare · Surplus electricity

3.1 Introduction

As we have seen in Chap. 2, in modeling feed-in tariffs (FITs), it is necessary to account for the distinctive features of investment in electricity generation from renewable energy sources. The most critical feature may be the fact that a decision is made about whether to invest in a renewable energy project, not about how much electricity to produce from renewable energy sources (RES-E).

The purpose of this chapter is to provide a new type of analytical model of FITs that includes the critical feature mentioned above. To develop as concrete a model as

Y. Yamamoto, *Feed-in Tariffs and the Economics of Renewable Energy*,
https://doi.org/10.1007/978-3-319-76864-9_3

possible, we consider a situation where investment in solar photovoltaic (PV) power generation is considered for both the business sector and the residential sector.

The model to be developed is partly based on the models by Yamamoto (2012) and Yamamoto (2017). Both of the models consider FITs for the adoption of PV systems in the residential sector. Yamamoto (2012) compares three types of FITs with regard to which part of PV electricity is eligible for FITs: all PV electricity, surplus PV electricity, and the PV electricity generated less the electricity consumption in a certain period. On the other hand, Yamamoto (2017) investigates a combination of FITs and capital subsidies. In contrast, the present model considers the business sector as well as the residential sector, although the model architecture is similar to those previous models.

The most important aspect of our model is that it considers fixed investment cost rather than variable operating cost. As mentioned above, in decision-making with regard to RES-E generation, whether to invest is much more important than how much to generate. In modeling this aspect, we need to explicitly describe a decision-maker's calculation about the investment.

The model can be briefly explained as follows. First, consider the business sector. If a firm invests in PV generation at a site, it gains revenue by selling PV electricity but incurs investment costs. The amount of PV electricity generated and the investment cost will vary among sites. Hence, given a certain price of PV electricity, some sites will be invested and others will not. Next, consider the residential sector. A household will decide on a PV system subject to income constraints. However, if a certain form of utility function is assumed, as is commonly done in the economics literature, a household's decision-making may be described similarly to that of a firm.

The remainder of the chapter is organized as follows. In Sect. 3.2, we develop a model of FITs for the business sector. Then, in Sect. 3.3, we examine the amount of PV electricity generated and the problem of social welfare maximization. In Sect. 3.4, a model of FITs for the residential sector is developed. While a firm makes a decision on investment in PV generation by comparing benefits and costs, a household maximizes utility subject to income constraints. The amount of PV electricity and the social welfare maximization problem are again addressed for the residential sector in Sect. 3.5. Section 3.6 concludes the chapter with a suggestion regarding the modeling conducted in the following chapters.

3.2 The Model for the Business Sector

In this section, we develop a model to investigate FITs for the business sector. First, we define variables that appear in the model, and then we examine the decision-making of a firm regarding whether to invest in PV generation at a site.

Table 3.1 Notations

I_i	i's cost of PV generation
v_i	Household i's satisfaction obtained from adopting a PV system
z_i	i's amount of PV electricity at site i or by household i
m_i	i's PV capacity installed
b	Environmental benefit per unit of PV electricity
c	Unit cost of conventional electricity generation
p	Price of PV electricity

3.2.1 Definition of Variables

Let us define the variables that appear in the model, which are summarized in Table 3.1.

Assume that a firm is considering investing in PV generation, which will operate during the lifetime of a PV system. The firm makes a decision based on the maximization of profit.

There are many potential sites in a country where a firm may install PV systems. Each site has various degrees of potential for PV generation depending on physical conditions such as ground area, solar radiation, and connection to the power grid. A site is denoted by i.

Suppose that site i is characterized by three parameters with respect to investment in PV generation. The first parameter is installed capacity, m_i, of PV panels, expressed in kW. For the sake of simplicity, m_i is expressed as an integer. The total amount of PV capacity throughout the country is $\sum_{i \in \Omega} m_i$, where Ω denotes the set of sites to be developed in the current term.

The second parameter is the amount of PV electricity, z_i, that a unit of PV capacity generates during the lifetime. For simplicity, it is assumed that z_i is a definite value expressed without probability, because, in practice, a firm can collect information on z_i by means of preliminary surveys. A discount factor is taken into account so that the firm can obtain the payment pm_iz_i over the PV system's lifetime if a unit of PV electricity is sold at a price p (Appendix 1).

The third parameter is the cost, I_i, of PV investment per unit capacity of the PV panel. This parameter includes the costs of equipment and installation, maintenance costs, ground rent, and so forth. All of the costs incurred beyond the current term are expressed in discounted presented values. Thus, the total cost is m_iI_i for the investment in PV generation at site i.

A unit cost of conventional power generation is denoted simply by c. It is presumed that c is also a single standard electricity retail rate.

Let b represent the benefit of the avoided external costs realized per unit of PV electricity (Klein 2008, pp. 16–17). This includes the benefits of climate change mitigation, energy security improvement, and so forth. Note that b does not include conventional generation cost c, which is also to be avoided.

A government promoting PV generation provides FITs: PV electricity is sold at a price, p, per kWh during a set period of years. For the sake of analytical simplicity, it is assumed that the period during which FITs are applied equals the lifetime of a PV system. Accordingly, if a firm invests in PV generation at site i, the investment yields revenue of $m_i p z_i$ and incurs a cost of $m_i I_i$.

3.2.2 *Decision-Making of a Firm*

A firm invests in PV generation at site i if and only if

$$p z_i - I_i \geq 0. \tag{3.1}$$

If p is offered by the government, site i is more likely to be developed as I_i is smaller or z_i is larger. Define Ω_p as the set of sites developed when p is offered: $\Omega_p = \{i \mid p z_i - I_i \geq 0\}$. Once site i has been developed, an installed capacity of m_i kW generates $m_i z_i$ kWh during the lifetime of the site.

It is obvious that if a government sets a higher price of p to attain a larger quantity of PV capacity or a larger quantity of PV electricity to be generated, a higher cost of FITs results. In other words, if the government is seeking to reduce the cost of FITs, the amount of PV electricity to be generated will become small.

3.3 Social Welfare Maximization for the Business Sector

As noted at the end of Sect. 3.2, there is a trade-off between increased PV electricity to be generated and a reduction in FIT costs. Accordingly, the amount of PV electricity to be generated will be small if a government is seeking to reduce the cost of FITs.

Let us examine another criterion concerning the level at which p should be set. Social welfare maximization is commonly examined in the economics literature in order to assess economic efficiency. It consists of the surplus or net benefit to market participants, typically consumers and producers, in equilibrium (Prima et al. 2011).

Let us define social welfare in the model. It may be defined as the sum of the profit to firms that invest in PV generation and the social benefit resulting from using PV electricity instead of conventional electricity generated with fossil fuels and nuclear power. Note that the cost to an electric utility of purchasing PV electricity does not appear as part of social welfare: it is passed on either to electricity customers through a higher electricity retail rate or to taxpayers through taxation; thus, the FIT cost to the utility and the burden on ratepayers or taxpayers cancel each other out.

Accordingly, social welfare is computed as follows. Suppose that a firm invests in PV generation at site i. The PV system at site i yields a profit, $p m_i z_i - m_i I_i$, to the firm, and it yields external benefits, $bm_i z_i$, to society. At the same time, the electric utility can save on the costs of conventional electricity generation, $cm_i z_i$. On the other hand, the cost of purchasing PV electricity, $pm_i z_i$, is incurred as either an electricity surcharge or a tax increase. Hence, the social welfare amounts to $\sum_{i \in \Omega_p} m_i \left[(b + c)z_i - I_i\right]$. Therefore, the social welfare maximization problem is

$$\text{maximize} \sum_{i \in \Omega_p} m_i \left[(b + c)z_i - I_i\right]. \tag{3.2}$$

The solution to Problem (3.2) is $p = b + c$. As noted in Sect. 3.1, the number of sites to be developed increases as p is set higher. However, from Eq. (3.1), if $p \leq b + c, (b + c)z_i - I_i \geq 0$ for $i \in \Omega_p$; if $p > b + c, (b + c)z_i - I_i < 0$ for $i \in \Omega_p \backslash \Omega_{b+c} = \left\{i \,|\, i \in \Omega_p, i \notin \Omega_{b+c}\right\}$. Therefore, the social welfare is maximized at $p = b + c$. Note that any p close to $b + c$ yields the same maximum as long as $\Omega_p = \Omega_{b+c}$.

In summary, to maximize social welfare, the FIT level should be set at the avoided cost per unit of PV electricity, that is, an avoided unit cost of conventional power generation plus the avoided external cost realized per unit of PV electricity. Note, in this case, that while the government succeeds at social welfare maximization, it cannot control the amount of PV electricity to be generated; that amount is determined endogenously. In other words, it would not be possible to simultaneously achieve both social welfare maximization and a target amount of PV electricity to be generated.

3.4 The Model for the Residential Sector

The model for the residential sector may be developed in essentially the same way as that for the business sector. However, there are three distinctive features that must be accounted for in modeling the decision-making of a household. Focusing on these features, let us develop the model for the residential sector.

First, the index i, which indicated a site where a firm installs a PV system in the model for the business sector, corresponds to each household considering adopting a PV system. The reason is that the site where a household installs a PV system is typically the rooftop, whereas a firm may have many potential sites to develop for PV generation.

Hence, m_i represents the PV capacity a household installs on the rooftop. Recall that in the model for the business sector, this term was the PV capacity a firm would install at site i. Note that because the area of the rooftop on which an adopter can install PV panels does not vary substantially among adopters, we can assume m_i has a fixed value, say 4 kW, regardless of adopter i, although we do not make such an assumption in this chapter.

Second, it may be helpful to incorporate an adopter's satisfaction into the utility a household gains from the adoption. This is based on the observation that some households adopt PV systems even if the adoption does not pay in a purely economic way. It is postulated that such households must derive various forms of satisfaction from the PV adoption that go beyond economic considerations. These include the good feelings of contributing to greenhouse gas emissions reduction, using an innovative technology, and generating electricity on their own. It is assumed that the household's satisfaction depends on the PV capacity it installs, and we let v_i represent this value per unit of PV capacity; that is, household i obtains satisfaction $m_i v_i$ from the adoption of a PV system.

The quantity of v_i may be evaluated by, for example, a contingent valuation method, thereby assessing willingness to pay (WTP) for a PV system. Specifically, we may conduct a questionnaire survey to question households about WTP, from which we subtract their expected $pm_i z_i$ to yield $m_i v_i$, assuming that their utility consists of $m_i v_i$ and $pm_i z_i$. In particular, if a household perceives no satisfaction, i.e., $v_i = 0$, the utility it obtains is $pm_i z_i$.

It should be noted that v_i and b are totally different notions each other. On the one hand, v_i is linked to an adopter, such that it partly forms the demand for a PV system. On the other hand, b is linked to society independently of the supply and demand for a PV system. In other words, while v_i is a notion considered within the market for a PV system, b is a notion considered outside the market.

Lastly, as is commonly posited in the economics literature, a household makes a decision about the consumption of a bundle of goods and services by maximizing utility subject to income constraints. A PV system may constitute that consumption bundle. Recall that a firm makes a decision on whether to invest in PV generation by comparing benefit with cost.

Accordingly, the decision-making process of a household may be modeled as follows. Let us begin by deducing a household i's demand function, $D_i\,(I_i)$, for a PV system. Supposing that there are only two types of goods, a PV system and a composite good, with a special form of utility function, which is commonly utilized in the economics literature, we may deduce a demand function as $D_i\,(I_i) = 0$ for $I_i > pz_i + v_i$ and $D_i\,(I_i) = 1$ for $I_i \le pz_i + v_i$ (Appendix 2).

In the model, it is assumed that an adopter sells all of the PV electricity it generates. This means that a household, on the one hand, purchases all of the electricity it consumes from an electric utility at a retail electricity rate, c; on the other hand, it sells all of the PV electricity it generates at a set price under FITs, p. The case where an adopter can sell only surplus PV electricity will be considered in Chaps. 4 and 5. It should be noted that FITs make sense even if $p < c$ because an adopter's PV generation is not always coincident with its electricity consumption.

Therefore, if a government offers a price for PV electricity in a FIT system, a household adopts a PV system if and only if

$$pz_i + v_i - I_i \ge 0. \tag{3.3}$$

As a result, the decision-making of a household is the same as that of a firm, i.e., Eq. (3.1), except for the term v_i. A household that has a larger z_i, larger v_i, or smaller I_i is more willing to adopt a PV system. If we define Ω_p as the set of adopters when p is offered as in the business sector, $\Omega_p = \{i \mid pz_i + v_i - I_i \geq 0\}$.

Obviously, as in the business sector, the government must set a higher p to make a larger number of households adopt PV systems or to make a larger amount of PV electricity be generated.

3.5 Social Welfare Maximization for the Residential Sector

Social welfare is defined as in Sect. 3.3: it consists of the utility all of the adopters gain and the social benefit less the cost of PV systems. Hence, the social welfare maximization problem is

$$\text{maximize} \sum_{i \in \Omega_p} m_i \left[(b + c)z_i + v_i - I_i\right]. \tag{3.4}$$

The solution to Problem (3.4) is $p = b + c$, which is the same as in the model for the business sector. To verify this, note from Eq. (3.3) that if $p \leq b + c$, $(b + c)z_i + v_i - I_i \geq 0$ for $i \in \Omega_p$; if $p > b + c$, $(b + c)z_i + v_i - I_i < 0$ for $i \in \Omega_p \backslash \Omega_{b+c} = \left\{i \mid i \in \Omega_p, i \notin \Omega_{b+c}\right\}$. Note that any p close to $b + c$ yields the same maximum as long as $\Omega_p = \Omega_{b+c}$.

In summary, two results are yielded by the model for the residential sector. First, if the government sets a higher price p to increase the amount of PV electricity generated, the cost of FITs, that is, the cost of purchasing PV electricity, increases as well. Second, social welfare is maximized at $p = b + c$ and in its neighborhood where p satisfies $\Omega_p = \Omega_{b+c}$. These are the same results as those for the business sector, although there are some differences in the decision-making processes.

3.6 Conclusion

In this chapter, we proposed a new way of modeling FITs for the business sector and the residential sector. Our approach sheds light on the fixed investment costs of PV generation rather than the variable operating costs. Furthermore, this approach considers the heterogeneity of investors, that is, firms or households, in order to explicitly describe their decision-making.

This chapter obtains the following results common to the business sector and the residential sector. First, the amount of PV electricity generated increases as the price of PV electricity is set higher. Second, social welfare is maximized at the PV electricity price that equals the avoided cost per unit of PV electricity. The avoided cost consists of the avoided cost of conventional power generation, such as coal and oil, and the external benefit produced by PV electricity, such as reduction of greenhouse gas emissions and improved energy security.

Some remarks should be made about the results. Generally, it will not be possible to achieve the two goals simultaneously, that is, generating the target amount of PV electricity and maximizing social welfare. Social welfare is maximized at the definite price mentioned just above, whereas the amount of PV electricity increases as the PV electricity price becomes higher. This may be attributable to the fact that a government has only a single control variable, a PV electricity price, while it has the two policy goals. Accordingly, a potential approach to reconciliation is to give the government a new control variable. This topic will be discussed in Chap. 5, where a government that seeks to maximize social welfare while simultaneously installing a target amount of PV capacity will grant an amount of capital subsidy in addition to offering FITs to investors in PV generation.

In the model for the residential sector, it was assumed that an adopter of a PV system can sell all of the PV electricity it generates at a set price under FITs. However, in reality, an adopter self-consumes some PV electricity and sells the surplus to an electric utility at the set price. In other words, FITs are applied not to all PV electricity but to surplus PV electricity. This contrasts sharply with the business sector, where all PV electricity generated is generally eligible for FITs. In the next chapter, we will investigate FITs for the residential sector in this regard.

Appendix 1: The Amount of PV Electricity to Be Generated

The assumption that a discount rate is taken into account may be mathematically paraphrased as follows. Let ζ_i denote site i's annual amount of PV electricity generated per unit of PV capacity. Then, the FIT payment that site i yields per unit of PV capacity throughout the period is $\sum_{t=1}^{T} p\zeta_i/(1+\varepsilon)^{t-1}$, where ε is a discount rate and T is the final year of the period. Our definition of z_i means $z_i = \sum_{t=1}^{T} \zeta_i/(1+\varepsilon)^{t-1}$ so that site i yields revenue $pm_i z_i$ throughout the period. It is remarked that $z_i = \zeta_i \left(1-\delta^T\right)/(1-\delta)$, where $\delta = 1/(1+\varepsilon)$.

Appendix 2: A Household's Demand Function

A household i's demand function for a PV system $D_i\,(I_i)$ is deduced as follows. Suppose there are only two types of goods: good X is a PV system, and good Y is a composite good, where $X = 1$ if a household adopts a PV system and $X = 0$ if it does not adopt. Let household i's utility and income be $U_i(X, Y)$ and M_i, respectively. In addition, let R_i be household i's reservation price for a PV system. The reservation price is the price at which a household is indifferent about whether to adopt a PV system. Thus, the demand function is described as $D_i\,(I_i) = 0$ if $I_i > R_i$ and $D_i\,(I_i) = 1$ if $I_i \leq R_i$. Hence, $U_i(0, M_i) = U_i(1, M_i - R_i)$ holds (Varian 2014).

For the sake of analytical simplicity, a special form of utility function, which is commonly utilized in the economics literature, is assumed: $U_i(X, Y) = pz_i X + Y$. Then, $M_i = pz_i + M_i - R_i$, or $R_i = pz_i$ follows from $U_i(0, M_i) = U_i(1, M_i - R_i)$. Plugging this into the demand function yields $D_i\ (I_i) = 0$ for $I_i > pz_i$ and $D_i\ (I_i) = 1$ for $I_i \leq pz_i$.

Note that the utility function allows us to disregard a household's income when examining its decision-making about the adoption of a PV system.

References

Klein A (2008) Feed-in tariff designs: options to support electricity generation from renewable energy sources. VDM Verlag Dr, Müller, Saarbrücken

Mehrdad Pirnia, Jatin Nathwani, David Fuller, (2011) Ontario Feed-in-Tariffs: System Planning Implications and Impacts on Social Welfare. The Electricity Journal 24 (8):18-28

Varian H (2014) Intermediate microeconomics, 9th edn. W.W. Norton & Company, New York

Yamamoto Y (2012) Pricing electricity from residential photovoltaic systems: a comparison of feed-in tariffs, net metering, and net purchase and sale. Sol Energy 86:2678–2685

Yamamoto Y (2017) Feed-in tariffs combined with capital subsidies for promoting the adoption of residential photovoltaic systems. Energy Policy 111:312–320

Chapter 4
Three Types of Feed-in Tariffs for the Residential Sector

Abstract To incentivize households to adopt a photovoltaic (PV) system, there are three types of feed-in tariffs with respect to what part of PV electricity is priced: all PV electricity, surplus PV electricity, and the difference between PV generation and electricity consumption. In this chapter, we refer to these as FITs for all PV electricity, FITs for surplus PV electricity, and net metering, respectively. This study aims to compare these mechanisms with respect to retail electricity rates, including the cost to an electric utility of purchasing PV electricity and with respect to social welfare. A simple microeconomic model is developed. The findings are as follows. First, the mechanism that yields the lowest surcharged electricity rate is not clear; it depends on the parameter values. If households are more homogeneous in terms of parameter values, the difference in the surcharged electricity rates will be small. Second, the mechanism that produces the largest social welfare is FITs for all PV electricity. These results are confirmed by means of a simulation. If we take account of some reductions in electricity consumption in the case of FITs for surplus PV electricity or net metering, the results for social welfare should be slightly modified. If the reduction is significantly large, FITs for surplus PV electricity or net metering may produce greater social welfare compared with FITs for all PV electricity.

Keywords Feed-in tariffs · Surplus electricity · Net metering
Photovoltaic system

4.1 Introduction

Residential electrical power generation using customer-owned, grid-connected, solar photovoltaic (PV) systems has become an attractive option for many households. While the appeal of PV systems can be attributed, in part, to technological devel-

Modified, with permission of Elesevier, from Yamamoto, Y., Pricing electricity from residential photovoltaic systems: a comparison of feed-in tariffs, net metering, and net purchase and sale, Solar Energy, 86, 2678–2685, Elsevier, 2012. I would like to thank Elsevier.

Y. Yamamoto, *Feed-in Tariffs and the Economics of Renewable Energy*,
https://doi.org/10.1007/978-3-319-76864-9_4

opments that have reduced the cost of PV power generation, feed-in tariffs (FITs) are also encouraging households to adopt PV systems. Generally, in a FIT system, adopters can sell PV electricity to an electric utility or distribution company at a set price for a set period of years. Although there are many variations on FITs, we look at three types of FITs with respect to the type of PV electricity that can be sold at a set price: all PV electricity, surplus PV electricity, and the difference between generation and consumption. In this chapter, we refer to these three types as FITs for all PV electricity, FITs for surplus PV electricity, and net metering, respectively.

We use the following definitions in this chapter, although different definitions may be possible for each of the three types of FITs. First, we define FITs for all PV electricity. Electric utilities are forced to purchase all of the PV electricity at a set price during a set number of years. The price and the period are determined by a government. This is equivalent to a situation in which adopters can sell all of the PV electricity they produce at a set price and must purchase all of the electricity they consume at standard electricity rates.

Next, we define FITs for surplus PV electricity. Electric utilities must purchase the electricity that adopters produce beyond their own consumption. In other words, only surplus PV electricity—the PV electricity actually fed into the grid at times when PV generation exceeds electricity consumption—is purchased at a set price. In this regard, the amounts of electricity generation and consumption are constantly compared, and an electric utility purchases the difference only when the former is larger than the latter.

Lastly, we define net metering. Whenever the PV generation of an adopter exceeds that adopter's electricity consumption, the electric meter runs backwards. At the end of a billing period, if the amount of PV generation is larger than that of electricity consumption, the adopter is paid for the net amount of PV generation at a set price; otherwise, the adopter must pay for the net amount of electricity consumption at standard electricity rates. Net metering differs from FITs for surplus PV electricity in that the amounts of PV generation and electricity consumption are compared at the end of every billing period under net metering, while under FITs for surplus PV electricity, the comparison is made moment-by-moment.

Many countries have implemented programs using these three mechanisms. Germany, for instance, adopted a program of FITs for all PV electricity; the German retail electricity rate was 26.14–27.81 €ct/kWh in 2012, while PV electricity was priced at 18.33–24.43 €ct/kWh, as reported on the websites of Europe's Energy Portal and the German Energy Blog. In contrast, Japan has adopted a program of FITs for surplus PV electricity (Mendonça 2007); Japan's retail electricity rate was ¥17.87–24.13 per kWh in 2012, while PV electricity was priced at ¥42.00 per kWh (Tokyo Electric Power Company). In the United States, according to the Database of State Incentives for Renewables and Efficiency (DSIRE) and Hughes and Bell (2006), many states use net metering; under these programs, the net amount of PV generation is often priced at the retail electricity rate.

In light of the above situations, the purpose of this chapter is to clarify what differences there may be among these three mechanisms. If the PV electricity price is equal to the retail electricity rate under each of the mechanisms, there is no difference among them (as shown below). However, if the PV electricity price and the retail electricity rate are different, then there may be differences among the mechanisms.

Prior works have investigated these mechanisms separately. For instance, a number of papers, including Menanteau et al. (2003), Mitchell et al. (2006), Fouquet and Johansson (2008), and Bürer and Wüstenhagen (2009), have compared FITs with a well-known, quantity-based support mechanism, the renewables portfolio standard (RPS). Many other studies, including Black (2004), Duke et al. (2005), Mills et al. (2008), Carley (2009), and Couture and Gagnon (2010), have investigated how net metering and FITs for surplus PV electricity are useful for promoting the adoption of renewable energy systems. To my knowledge, however, no previous studies have investigated the three mechanisms comparatively.

To address this gap in the research, this study compares the three mechanisms with respect to social welfare and retail electricity rates; retail electricity rates are increased so that the cost to an electric utility of purchasing PV electricity is fully transferred to all electricity consumers equally through rate increases. Furthermore, in the economics literature, social welfare is commonly used to examine economic efficiency. Social welfare is defined here as the sum of consumer surplus, profit for the electric utility, and environmental benefit. Note that PV generation will yield environmental benefits, for example, reductions in greenhouse gas emissions resulting from decreased use of the electricity generated from fossil fuels, as well as improvements in energy security owing to the use of solar radiation in a particular country.

This study makes the comparison theoretically by developing a simple microeconomic model. If a government implements a program with one of these mechanisms, how it should determine the PV electricity price? Suppose that the goal of the government is to increase PV generation in the residential sector. To this end, a potential approach would be for the government to set a PV electricity price under FITs, such that a certain number of households adopt PV systems. However, the PV electricity price necessary to achieve the target number of household adoptions may vary depending on the mechanism chosen. This problem may be addressed by a model investigation.

Our model pays attention to the heterogeneity among households in terms of electricity consumption, PV generation, PV system cost, and the utility obtained from the adoption of a PV system. It is assumed in the model that households decide whether to adopt a PV system by maximizing utility. Then, some households will adopt a PV system while others will not, depending on their own calculation of preference. Using this model, we characterize each mechanism and then compare the three mechanisms with respect to surcharged retail electricity rates and social welfare. A numerical example is also provided to aid the reader in understanding the results of the theoretical investigation.

The remainder of the chapter is organized as follows. In Sect. 4.2, the model is described. A mathematical representation of each of the three mechanisms is presented in Sect. 4.3. In Sect. 4.4, the mechanisms are compared with respect to surcharged electricity retail rates and social welfare. To help explain the theoretical results, Sect. 4.5 presents a numerical example. In Sect. 4.6, the effect of reduced electricity consumption is considered. Finally, Sect. 4.7 concludes the chapter.

Table 4.1 Notations

I_i	Household i's cost of a PV system
v_i	Household i satisfaction obtained from adopting a PV system
b	Benefit of the avoided external cost per unit of PV electricity
c	Unit cost of conventional power generation
r	Standard electricity retail rate
p	Price of PV electricity
z_i	Household i's amount of PV generation
$z_{g,i}$	Household i's PV generation in net generation periods
$z_{c,i}$	Household i's PV generation in net consumption periods
x_i	Household i's amount of surplus PV electricity
y_i	Household i's amount of self-consumed PV electricity
q_i	Household i's electricity consumption
$q_{g,i}$	Household i's electricity consumption in net generation periods
$q_{c,i}$	Household i's electricity consumption in net consumption periods

4.2 The Model

Let us set up a model to investigate three types of FITs. Considering that FITs are implemented to promote further development of the electricity generated from renewable energy sources (RES-E), we model a society with an earlier stage of diffusion of PV systems. This means that the model assumes the total amount of electricity consumption is much larger than the amount of PV electricity to be generated.

Key notations, which are summarized in Table 4.1, are essentially the same as in the model for the residential sector in Chap. 3. The only difference is that the PV capacity an adopter will install on a rooftop is fixed at a certain value in this chapter because it would not vary greatly among adopters. Accordingly, the values expressed per unit of PV capacity are substituted with the values expressed per household. The variables are described again for convenience, along with some new notations defined in this chapter.

Let N be the set of all households. The number of households in N is also represented by N to avoid having too many symbols in the model. A government aims to have n of N households adopt a PV system, either by means of FITs for all PV electricity, FITs for surplus PV electricity, or net metering.

For a household i, let I_i and v_i denote the cost of a PV system and the satisfaction household i experiences if it adopts a PV system, respectively.

Let c represent the cost to the electric utility of generating a unit of electricity from conventional energy sources; c is also the retail electricity rate before the implementation of FITs.

Let b denote the environmental benefit per unit of PV electricity, including reduction in greenhouse gas emission, improvement of energy security, and other benefits. Note that b does not include the avoided cost, c, of conventional electricity generation.

Let q_i denote the amount of household i's electricity consumption. It is assumed that q_i remains constant if retail electricity rates change. This assumption, which helps to keep the model simple, is reasonable because electricity is a necessity of daily life, such that the price elasticity of demand is significantly small. Define $L = \sum_{i \in N} q_i$.

Let z_i denote the amount of household i's electricity consumption. It is also assumed that $z_i < q_i$, in accordance with Duke et al. (2005), taking into account that a common residential PV system typically generates nearly half the average annual electricity consumption of a U.S. household.

To investigate FITs for surplus PV electricity and net metering, more detailed descriptions of z_i and q_i are needed. On the one hand, in the case of FITs for surplus PV electricity, the amount of PV electricity, z_i, is divided into x_i and y_i: x_i is the amount of PV electricity that is not self-consumed by household i but rather fed into the grid, and y_i is the amount actually self-consumed by i.

On the other hand, in the case of net metering, the balance of PV generation and electricity consumption matters. If a billing period is shorter than a year, as is often the case, household i's amount of PV generation may be larger than its electricity consumption during some billing periods, and it may be smaller in other periods. Billing periods in which the amount of PV generation is larger than that of electricity consumption are here termed *net generation periods*, and periods in which the opposite is the case are termed *net consumption periods*. Let $z_{g,i}$ denote the amount of PV electricity generated, and $q_{g,i}$ the amount of electricity consumption in net generation periods; $z_{c,i}$ and $q_{c,i}$ are similarly defined for net consumption periods. Thus, $z_{g,i} + z_{c,i} = z_i$, and $q_{g,i} + q_{c,i} = q_i$, where $z_{g,i} > q_{g,i}$, $z_{c,i} < q_{c,i}$.

For any of the three FIT mechanisms, the government sets the price of PV electricity at p and the retail electricity rate at r, where r is set in place of c so that the electric utility can fully cover the cost of purchasing PV electricity from adopters.

4.3 Mathematical Representations of the Mechanisms

This section presents mathematical representations for each of the three mechanisms. FITs for all PV electricity, FITs for surplus PV electricity, and net metering are dealt with in Sects. 4.3.1, 4.3.2, and 4.3.3, respectively.

To begin, recalling from Sect. 3.4 in Chap. 3 that household i's demand function for a PV system is $D_i(I_i) = 0$ for $I_i > pz_i + v_i$ and $D_i(I_i) = 1$ for $I_i \leq pz_i + v_i$, we assume that no households adopt a PV system under any mechanism if $p = c$, i.e., $v_i < I_i - cz_i$. This assumption is made for the sake of analytical simplicity and does not entail any loss of generality: suppose $p = c$. Then, $r = c$ follows. Accordingly, the set of adopters is identical between the three mechanisms because they are identical in this case. Hence, if there are adopters when $p = c$, we can simply subtract their number from n before the analysis.

4.3.1 FITs for All PV Electricity

Suppose that the government implements a program of FITs for all PV electricity, where adopters can sell at p all of the PV electricity they generate, while purchasing at r all of the electricity they consume. To be concrete, an adopter i sells z_i at p, while purchasing q_i at r. This implies that the cost of a PV system effectively decreases by pz_i so that household i adopts a PV system if and only if $v_i \geq I_i - pz_i$.

First, let us determine the PV electricity price at which n households adopt PV systems. Define $p_{F,i}$ as the smallest p that satisfies $v_i \geq I_i - pz_i$, i.e.,

$$p_{F,i} = \frac{I_i - v_i}{z_i}. \tag{4.1}$$

Thus, $p_{F,i}$ is the threshold price of PV electricity for household i to adopt a PV system: it adopts if and only if $p \geq p_{F,i}$. Note that $p_{F,i} > c$ due to the assumption $v_i < I_i - cz_i$, and note that $p_{F,i}$ is independent of r. Given a p, household i is more willing to adopt as z_i becomes large or $I_i - v_i$ becomes small.

Let p_F be the nth smallest $p_{F,i}$ of $\{p_{F,i} | i \in N\}$ (Fig. 4.1) and define a set F as $F = \{i | p_{F,i} \leq p_F\}$. Then, p_F is the price of PV electricity such that n households adopt a PV system. In other words, p_F is the smallest p the government can set to have n households adopt. Note that p_F is greater than c and independent of r.

Second, let us determine the surcharged electricity rate r at which the cost to the electric utility of purchasing PV electricity can be covered. Once FITs are implemented, the electric utility sells $\sum_{i \in N} q_i$ at r, generates $\sum_{i \in N} q_i - \sum_{i \in F} z_i$ with unit cost c, and purchases $\sum_{i \in F} z_i$ at p_F. Hence, the change in the utility's profit is $(r - c) \sum_{i \in N} q_i - (p_F - c) \sum_{i \in F} z_i$. Therefore,

$$(r - c) \sum_{i \in N} q_i - (p_F - c) \sum_{i \in F} z_i = 0 \tag{4.2}$$

must be satisfied. Let r_F denote the r that solves Eq. (4.2).

Lastly, let us calculate the change in social welfare. Recall that social welfare consists of consumer surplus, the utility's profit, and environmental benefit. First, calculate the change in consumer surplus of a household. Before FITs are implemented, a household i purchases q_i units of electricity at c. Once FITs are offered, an adopter i purchases q_i units of electricity at r_F, incurs cost I_i, gains satisfaction v_i, and sells z_i units of PV electricity at p_F. Hence, the change in the consumer surplus of adopter i is $-(r_F - c)q_i - I_i + v_i + p_F z_i$, or $-(r_F - c)\, q_i + \left(p_F - p_{F,i}\right) z_i$ from Eq. (4.1). The first term corresponds to the change in electricity rates, and the second term is the premium paid to i for the adoption. On the other hand, a non-adopter j

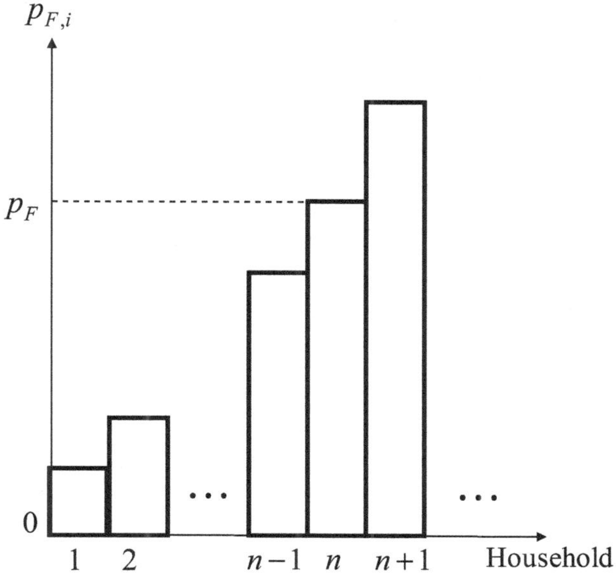

Fig. 4.1 A schematic illustration of how to determine p_F and F. Set the PV electricity price p_F at the nth smallest $p_{F,i}$. The set F of adopters consists of the households whose $p_{F,i} \leq p_F$

purchases q_j at r_F after FITs are implemented. The change in consumer surplus for j is therefore $-(r_F - c)q_j$. This is exactly attributable to the change in retail electricity rates.

Next, from Eq. (4.2), the change in the utility's profit is $(r_F - c)\sum_{i\in N} q_i - (p_F - c)\sum_{i\in F} z_i$. The first term corresponds to the change in revenue from electricity sales; the second term is divided into the premium paid to adopters, $\sum_{i\in F}\left(p_F - p_{F,i}\right) z_i$, and the social cost of having n households adopt, $\sum_{i\in F}\left(p_{F,i} - c\right) z_i$. Note that the profit is zero from Eq. (4.2) since $r = r_F$.

The environmental benefit resulting from the installation of PV systems is $b\sum_{i\in F} z_i$.

Hence, the change in social welfare, which is the sum of the changes in the sum of the consumer surplus, the profit of the electric utility, and the environmental benefit, is yielded by using Eq. (4.1) as

$$\sum_{i\in F} [(b + c)z_i + v_i - I_i]. \tag{4.3}$$

This makes intuitive sense: once FITs are offered, electricity amounting to $\sum_{i\in F} z_i$ is generated by PV systems at zero marginal cost, which was previously generated by the electric utility at unit cost c. In addition, the adoption of a PV system provides utility v_i and incurs cost I_i to adopter i, and it contributes $b\sum_{i\in F} z_i$ to society.

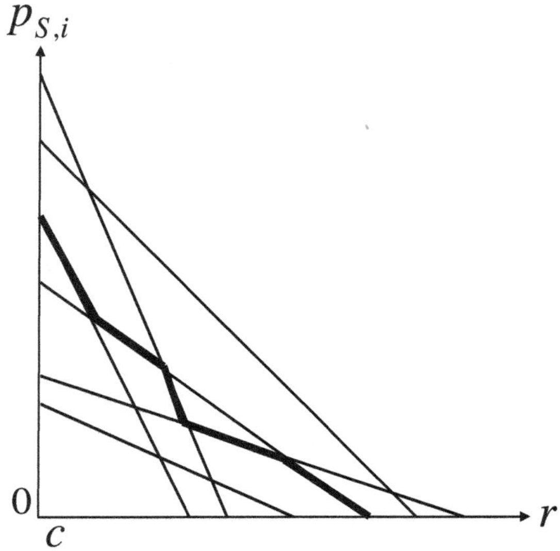

Fig. 4.2 A schematic illustration of how to determine $p_S(r)$ and $S(r)$. A downward-sloping line $p_{S,i}(r)$ represents household i's threshold price of PV electricity. For any r, set $p_S(r)$ at the nth smallest $p_{S,i}(r)$. The broken bold line is $p_S(r)$, shown for the case $n = 4$ for illustration. The set $S(r)$ of adopters consists of the households whose $p_{S,i}(r) \leq p_S(r)$

4.3.2 FITs for Surplus PV Electricity

Suppose that the government implements a program of FITs for surplus PV electricity, where adopters can sell surplus PV electricity instead of all PV electricity. An adopter sells x_i and self-consumes y_i units of PV electricity. In other words, it sells x_i at p and purchases $q_i - y_i$ at r. This is equivalent to the situation in which i purchases q_i at r and sells x_i at p and y_i at r. Accordingly, the cost of a PV system decreases by $px_i + ry_i$ for i. Hence, household i adopts a PV system if and only if $v_i \geq I_i - px_i - ry_i$.

First, let us work out the PV electricity price. Define $p_{S,i}(r)$ as the smallest p at which household i would adopt, i.e.,

$$p_{S,i}(r) = \frac{I_i - v_i - ry_i}{x_i}. \tag{4.4}$$

Thus, $p_{S,i}(r)$ is the threshold price of PV electricity for household i to adopt a PV system: it adopts if and only if $p \geq p_{S,i}$. It is a linearly decreasing function of r (Fig. 4.2). Note that as x_i or y_i becomes large, or $I_i - v_i$ becomes small, household i becomes more willing to adopt a PV system.

Similarly to p_F and F, $p_S(r)$ and $S(r)$ are defined: $p_S(r)$ is the nth smallest $p_{S,i}(r)$ of $\{p_{S,i}(r) | i \in N\}$ (Fig. 4.2) and $S(r) = \{i | p_{S,i}(r) \leq p_S(r)\}$. Notice that $p_S(r)$ is the price of PV electricity the government sets to have n households adopt PV systems. It is a strictly downward-sloping, broken line in the *rp*–plane. Furthermore, $S(r)$ is the set of adopters, which generally differs from F.

Second, let us determine the surcharged electricity rate r at which the cost to the electric utility of purchasing PV electricity can be covered. The utility effectively sells $\sum_{i \in N} q_i$ at r, generates $\sum_{i \in N} q_i - \sum_{i \in S(r)} z_i$ at unit cost c, and purchases $\sum_{i \in S(r)} x_i$ at $p_S(r)$ and $\sum_{i \in S(r)} y_i$ at r. Thus, the change in the utility's profit is $(r - c) \sum_{i \in N} q_i - \sum_{i \in S(r)} (p_S(r)x_i + ry_i - cz_i)$. Therefore,

$$(r - c) \sum_{i \in N} q_i - \sum_{i \in S(r)} (p_S(r)x_i + ry_i - cz_i) = 0 \tag{4.5}$$

must be satisfied. Let r_S denote the r that solves Eq. (4.5). Define $p_S = p_S(r_S)$ and $S = S(r_S)$.

Lastly, we calculate the change in social welfare. First, compute the change in consumer surplus. An adopter i effectively purchases q_i at r_S, incurs cost I_i, gains satisfaction v_i, and sells x_i at p_S and y_i at r_S. On the other hand, non-adopter j purchases q_j at r_S. Accordingly, the changes in consumer surplus of adopter i and adopter j are $-(r_S - c)q_i + (p_S - p_{S,i}(r_S))x_i$ and $-(r_S - c)q_j$, respectively, by using Eq. (4.4). The change in the electric utility's profit is $(r_S - c) \sum_{i \in N} q_i - \sum_{i \in S} (p_S x_i + r_S y_i - cz_i)$. Hence, the change in social welfare is the sum of these and the environmental benefit $b \sum_{i \in S} z_i$ and it amounts to

$$\sum_{i \in S} [(b + c)z_i + v_i - I_i]. \tag{4.6}$$

It should be noticed that the social welfare is the same as in the case of FITs for all PV electricity, except for the set of adopters.

4.3.3 Net Metering

Suppose that the government implements net metering, where the balance between PV generation and electricity consumption is priced at the end of a billing period. An adopter i sells at p the net amount of PV electricity it generates, $z_{g,i} - q_{g,i}$, during net generation periods, and it purchases at r the net amount of electricity it consumes, $q_{c,i} - z_{c,i}$, during net consumption periods. Because $q_{g,i} + q_{c,i} = q_i$, this is equivalent to the situation where it sells $z_{g,i} - q_{g,i}$ at p and $z_{c,i} + q_{g,i}$ at r, while purchasing q_i at r. Accordingly, the price of a PV system decreases by $p(z_{g,i} - q_{g,i}) + r(z_{c,i} + q_{g,i})$ for household i. Thus, it would adopt a PV system if and only if $v_i \geq I_i - p(z_{g,i} - q_{g,i}) - r(z_{c,i} + q_{g,i})$.

First, let us calculate the PV electricity price. Define $p_{M,i}(r)$ as the smallest p at which household i would adopt, i.e.,

$$p_{M,i}(r) = \frac{I_i - v_i - r(z_{c,i} + q_{g,i})}{z_{g,i} - q_{g,i}}. \tag{4.7}$$

Thus, $p_{M,i}(r)$ is the threshold price of PV electricity for i to adopt a PV system; it is a linearly decreasing function of r. Note that as $z_{g,i} - q_{g,i}$ or $z_{c,i} + q_{g,i}$ becomes large, or as $I_i - v_i$ becomes small, household i becomes more willing to adopt.

As before, $p_M(r)$ and $M(r)$ are defined as follows: $p_M(r)$ is the nth smallest $p_{M,i}(r)$ of $\{p_{M,i}(r)|i \in N\}$ and $M(r) = \{i|p_{M,i}(r) \leq p_M(r)\}$. Notice that $p_M(r)$ is the price of PV electricity the government sets to have n households adopt PV systems. It is a strictly downward-sloping, broken line in the rp–plane. Furthermore, $M(r)$ is the set of adopters.

Second, let us determine the surcharged electricity rate r. The electric utility effectively sells $\sum_{i \in N} q_i$ at r, generates $\sum_{i \in N} q_i - \sum_{i \in M(r)} z_i$ with unit cost c, purchases $\sum_{i \in M(r)} (z_{g,i} - q_{g,i})$ at $p_M(r)$ and $\sum_{i \in M(r)} (z_{c,i} + q_{g,i})$ at r. Hence, the change in the utility's profit is $(r-c)\sum_{i \in N} q_i - \sum_{i \in M(r)} \left[p_M(r)(z_{g,i} - q_{g,i}) + r(z_{c,i} + q_{g,i}) - cz_i\right]$. Therefore,

$$(r - c)\sum_{i \in N} q_i - \sum_{i \in M(r)} \left[p_M(r)(z_{g,i} - q_{g,i}) + r(z_{c,i} + q_{g,i}) - cz_i\right] = 0 \quad (4.8)$$

must be satisfied. Let r_M denote the r that solves Eq. (4.8). Define $p_M = p_M(r_M)$ and $M = M(r_M)$.

Lastly, let us work out the change in social welfare. First, calculate the change in consumer surplus. An adopter i effectively purchases q_i at r_M, incurs cost I_i, gains satisfaction v_i, and sells $z_{g,i} - q_{g,i}$ at p_M and $z_{c,i} + q_{g,i}$ at r_M. On the other hand, non-adopter j purchases q_j at r_M. Accordingly, the changes in consumer surplus of adopter i and non-adopter j are $-(r_M - c)q_i + (p_M - p_{M,i}(r_M))(z_{g,i} - q_{g,i})$ and $-(r_M - c)q_j$, respectively, using Eq. (4.7). The change in the utility's profit is $(r_M - c)\sum_{i \in N} q_i - \sum_{i \in M} \left[p_M(z_{g,i} - q_{g,i}) + r_M(z_{c,i} + q_{g,i}) - cz_i\right]$. Hence, the change in social welfare is the sum of these and the environmental benefit $b\sum_{i \in M} z_i$, resulting in

$$\sum_{i \in M} [(b + c)z_i + v_i - I_i]. \quad (4.9)$$

It should be noted that net metering is identical to FITs for surplus PV electricity except that, in addition to the difference between the sets of adopters, x_i and y_i in FITs for surplus PV electricity are substituted with $z_{g,i} - q_{g,i}$ and $z_{c,i} + q_{g,i}$, respectively, in net metering.

4.4 Comparison of the Mechanisms

This section offers comparisons of the mechanisms with respect to, first, which yields the lowest electricity rate when the cost to the electric utility of purchasing PV electricity is passed on to all customers, and next, which mechanism produces the greatest social welfare.

4.4.1 Surcharged Electricity Rates

All customers of the electric utility are surcharged the cost to the utility of purchasing PV electricity. We have defined the surcharged electricity rates in Sect. 4.3 as r_F, r_S, and r_M associated with FITs for all PV electricity, FITs for surplus PV electricity, and net metering, respectively. We investigate which yields the lowest surcharged electricity rates, r_F, r_S, or r_M.

First, let us compare r_F and r_S. From Eq. (4.2), r_F is the r at the intersection of the two lines, $t = \alpha(r)$ and $t = \varphi(r)$, in the rt–plane, where $\alpha(r) = (r - c)\sum_{i \in N} q_i$ and $\varphi(r) = (p_F - c)\sum_{i \in F} z_i$. We can see that $t = \alpha(r)$ is an upward-sloping line, while $t = \varphi(r)$ is a horizontal line (Fig. 4.3). It is remarked that $r_F < p_F$ because generally $\sum_{i \in N} q_i > \sum_{i \in F} z_i$ in Eq. (4.2).

Similarly, from Eq. (4.5), r_S is the r at the intersection of the two curves, $t = \alpha(r)$ and $t = \psi(r)$ in the rt–plane, where $\psi(r) = \sum_{i \in S(r)} (p_S(r)x_i + ry_i - cz_i)$. Curve $t = \psi(r)$ is a broken line (Fig. 4.3): it is a straight line in each segment of r in which $S(r)$ and $p_S(r)$ remain unchanged. Two remarks should be made with regard to $\psi(r)$. First, differentiating $\psi(r)$ with respect to r in any one of the segments indicates that the slope is $\sum_{i \in S(r)} (y_i - x_i y_m / x_m)$, where subscript m represents the marginal household, the household whose threshold price equals $p_S(r)$, in that segment. Whether the slope is positive or negative may not be clear because it is not clear which is larger, y_i/x_i or y_m/x_m. Second, because $p_S(p_F) = p_F$ and $S(p_F) = F$ (Appendix 1), $t = \psi(r)$ intersects with $t = \varphi(r)$ when $r = p_F$ (Fig. 4.3).

Hence, it is not clear whether the intersection of $t = \alpha(r)$ and $t = \psi(r)$ is above or below $t = \varphi(r)$ in the rt–plane (see Fig. 4.3). If the intersection is below $t = \varphi(r)$, $r_F > r_S$ follows (Fig. 4.3a), and vice versa (Fig. 4.3b). Therefore, it is not clear whether $r_F > r_S$ or $r_F < r_S$.

Note that if households are more homogeneous, the difference between r_F and r_S may be smaller. Suppose there are few differences among households in terms of the relevant parameters. Then, $t = \psi(r)$ comes close to $t = \varphi(r)$ in Fig. 4.3, and hence r_S may be close to r_F.

Second, compare the surcharged electricity rates between FITs for all PV electricity and net metering, that is, between r_F and r_M. A similar investigation may indicate that whether $r_F > r_M$ or $r_F < r_M$ is not clear. If households are more homogeneous, the difference between r_F and r_M may be smaller.

4.4.2 Social Welfare

Among the three mechanisms, we compare the amount of social welfare. As shown in Sect. 4.3, social welfare has the same form regardless of the mechanism employed, i.e., $\sum [(b + c)z_i + v_i - I_i]$, where the summation is taken over all adopters.

Let A denote any one of the mechanisms. It also represents the set of adopters under mechanism A. Let $p_{A,i}$ be the threshold price of PV electricity for a household

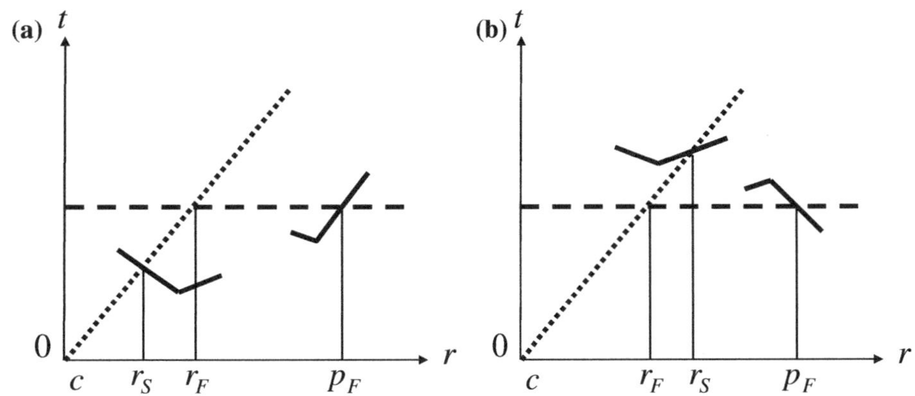

Fig. 4.3 Examples of the configuration of $t = \alpha(r)$, $t = \varphi(r)$, and $t = \psi(r)$. The dotted line, the dashed line, and the solid broken line represent $t = \alpha(r)$, $t = \varphi(r)$, and $t = \psi(r)$, respectively. The intersection of $t = \alpha(r)$ and $t = \varphi(r)$ yields r_F; the intersection of $t = \alpha(r)$ and $t = \psi(r)$ yields r_S. On the one hand, **a** if the intersection of $t = \alpha(r)$ and $t = \psi(r)$ is below $t = \varphi(r)$, $r_S < r_F$ follows. On the other hand, **b** if the intersection of $t = \alpha(r)$ and $t = \psi(r)$ is above $t = \varphi(r)$, $r_F < r_S$ follows

i to adopt a PV system, and p_A the PV electricity price for exactly n households to adopt PV systems.

First, we compare the social welfare between FITs for all PV electricity and FITs for surplus PV electricity. To this end, we determine which set, $\{p_{F,i}|i \in N\}$ or $\{p_{S,i}(r_S)|i \in N\}$, can more accurately place the elements of $\{I_i - v_i - (b+c)z_i|i \in N\}$ in ascending order if the elements of $\{p_{F,i}|i \in N\}$ and $\{p_{S,i}(r_S)|i \in N\}$ are individually arranged in ascending order. Recall that p_A is the nth smallest $p_{A,i}$, such that any household i having $p_{A,i} \le p_A$ adopts a PV system. Thus, put into reverse, when the elements of $\{p_{A,i}|i \in N\}$ are arranged in ascending order, mechanism A minimizes $\sum_{i \in A}[I_i - v_i - (b+c)z_i]$, or maximizes $\sum_{i \in A}[(b+c)z_i + v_i - I_i]$, if the elements of $\{I_i - v_i - (b+c)z_i|i \in N\}$ are also in ascending order.

To strictly examine the above ordering may be somewhat complicated, but considering it intuitively may offer sufficient clarity. The values of $I_i - v_i - (b+c)z_i$ and $p_{F,i} = (I_i - v_i)/z_i$ are both determined by only $I_i - v_i$ and z_i. On the other hand, $p_{S,i}(r_S) = [I_i - v_i - r_S(z_i - x_i)]/x_i$ is not determined in this manner; the ingredient of z_i, i.e., x_i, also matters. Therefore, $\{p_{F,i}\}$ may be more consistent with $\{I_i - v_i - (b+c)z_i\}$ in terms of ordering than $\{p_{S,i}(r)\}$. Hence, FITs for all PV electricity are likely to produce larger social welfare than FITs for surplus PV electricity.

The comparison may similarly be made between FITs for all PV electricity and net metering. Note that $P_{M,i}(r_M) = [I_i - v_i - r_M(z_i - z_{g,i} + q_{g,i})]/(z_{g,i} - q_{g,i})$ is not determined by only $I_i - v_i$ and z_i: $z_{g,i} - q_{g,i}$ matters. Therefore, $\{p_{F,i}\}$ may be more consistent with $\{I_i - v_i - (b+c)z_i\}$ in terms of ordering than $\{p_{M,i}(r_M)\}$. Hence, FITs for all PV electricity are likely to produce greater social welfare than is net metering.

In summary, FITs for all PV electricity produce greater social welfare than do FITs for surplus PV electricity and than does net metering.

As a corollary of the analysis above, an alternative mechanism, which produces larger social welfare than FITs for all PV electricity, may be possible (Appendix 2).

4.5 A Numerical Example

In this section, a numerical example is presented to supplement the results in Sect. 4.4. However, this should not be interpreted as an empirical investigation, which is beyond our present scope.

First, parameter values are set, and then simulation results are presented with discussion.

4.5.1 Setting of Parameter Values

Parameter values are set based on several sources (Table 4.2): the conventional power generation cost c = \$0.11/kWh (EIA 2017); the social benefit per

Table 4.2 Parameter values

I_i	Household i's cost of a PV system	Drawn from a normal distribution (mean: \$14,000, SD: \$1000)
v_i	Household i's satisfaction obtained from adopting a PV system	Drawn from a normal distribution (mean: \$500, SD: \$200)
z_i	Household i's amount of PV generation	Drawn from a normal distribution (mean: 80,000 kWh, SD: 16,000 kWh)
x_i	Household i's amount of surplus PV electricity	$z_i - y_i$
y_i	Household i's amount of self-consumed PV electricity	z_i multiplied by a random variable drawn from a uniform distribution ranging from 0.2 to 0.4
b	Benefit of the avoided external cost per unit of PV electricity	\$0.01/kWh
c	Unit cost of conventional power generation	\$0.11/kWh
N	Total number of households	10,000
n	Target number of adopters	100
L	Electricity consumption by all households	1060 GWh
T	Lifetime of a PV system	20 years
ε	Discount rate	0.01

unit of PV electricity $b = \$0.01/\text{kWh}$ in terms of CO_2 emissions reduction ($0.5\text{kgCO}_2/\text{kWh} \times \$\,20/\text{tCO}_2$: JPEA 2015; IEA 2012); and the fixed sum of electricity consumption $L = 1060\,\text{GWh}$ during the lifetime of a PV system $T = 20$ years ($5300\,\text{kWh/year} \times 10,000$ households $\times$ 20 years: JPEA 2015).

It is presumed that a PV system cost, I_i, is drawn from a normal distribution, with the mean of \$14,000 (METI 2013) and standard deviation of \$1000.

It is not easy to set v_i, a household's valuation of a PV system; this type of data will not generally be available, meaning that we need to conduct, for example, a questionnaire survey to estimate v_i. However, such an estimation is beyond the scope of this chapter, even though it would be an interesting research topic. Instead, we set that value artificially: it is assumed that v_i is drawn from a normal distribution with a mean of \$500 and a standard deviation of \$200. Recall that we have introduced v_i because some households adopt a PV system despite the fact that the adoption is less likely to pay. We may typically consider v_i much smaller than I_i.

It is presumed that household i's amount of PV electricity, z_i, to be generated during the lifetime of a PV system, is drawn from a normal distribution, with a mean of 80,000 kWh ($4400\,\text{kWh/year} \times (1 - 0.99^{20})/(1 - 0.99)$: NEDO 2017) and a standard deviation of 16,000 kWh. In this calculation, we set a discount rate of $\varepsilon = 0.01$. Assuming that each household consumes, on average, 30% of the PV electricity it generates, i's self-consumption of electricity, y_i, is constructed from z_i

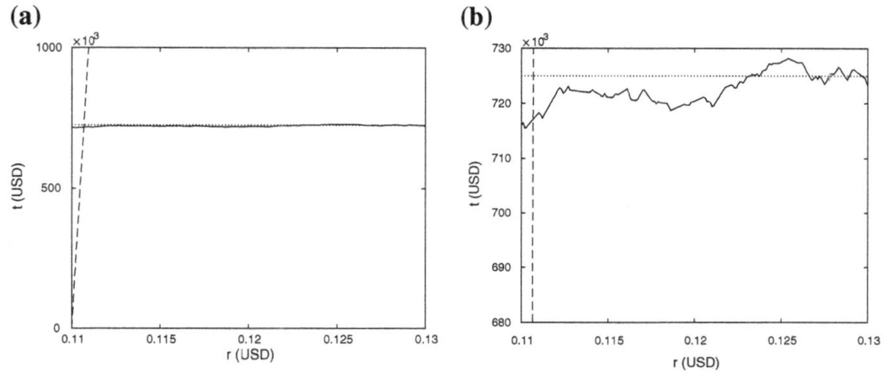

Fig. 4.4 A simulation result of $t = \alpha(r)$, $t = \varphi(r)$, and $t = \psi(r)$. The dotted line, the dashed line, and the solid broken line represent $t = \alpha(r)$, $t = \varphi(r)$, and $t = \psi(r)$, respectively. The solid curve $t = \psi(r)$ seems almost horizontal in (**a**), but a close look in (**b**) reveals that its intersection with the dashed line $t = \varphi(r)$ is below the dotted line $t = \alpha(r)$. Thus, $r_F > r_S$ results

multiplied by a random variable drawn from a uniform distribution ranging from 0.2 to 0.4. Finally, the surplus electricity, x_i, is computed as $x_i = z_i - y_i$.

The target number of adopters is $n = 500$ in relation to the total number of households (detached houses), $N = 10,000$. These values may be viewed as being expressed in, for example, thousands.

4.5.2 Simulation Results and Discussion

Simulations are conducted to compare FITs for all PV electricity and FITs for surplus PV electricity. As was examined in Sects. 4.3 and 4.4, net metering is essentially the same as FITs for surplus PV electricity in terms of surcharged electricity rates and social welfare maximization. Thus, our simulation focuses on FITs for all PV electricity and FITs for surplus PV electricity.

First, we compare surcharged electricity rates between r_F and r_S. Recall that r_F corresponds to the intersection of $t = \alpha(r)$ and $t = \varphi(r)$, and r_S corresponds to the intersection of $t = \alpha(r)$ and $t = \psi(r)$. We obtain two types of results with respect to whether the intersection of $t = \alpha(r)$ and $t = \psi(r)$ is below or above $t = \varphi(r)$ (Figs. 4.4 and 4.5). If the intersection is below $t = \varphi(r)$, $r_F > r_S$ holds (Fig. 4.4), and vice versa (Fig. 4.5). These results match those yielded theoretically in Sect. 4.4.1.

However, the simulation results do not reveal any differences between the values; $r_F = r_S = 0.1107$ in both Figs. 4.4 and 4.5. This is attributable to the assumption that we consider an earlier stage of diffusion, where the total amount of electricity consumption L is substantially larger than the amount of PV generation; under this assumption, $r - c \approx 0$ holds from Eqs. (4.2), (4.5), and (4.8) regardless of the mech-

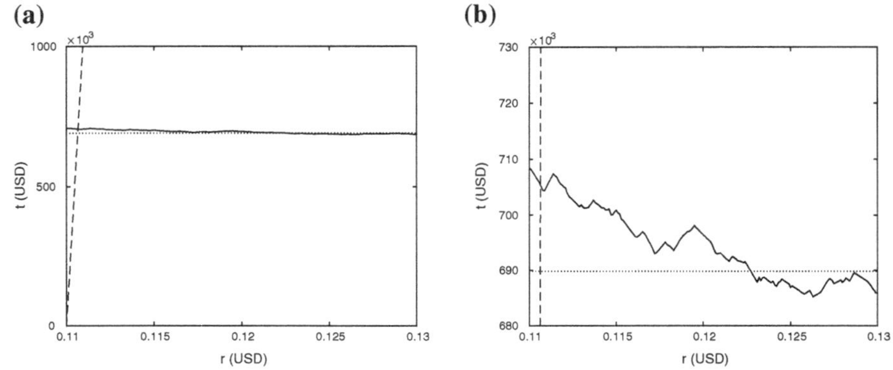

Fig. 4.5 Another simulation result of $t = \alpha(r)$, $t = \varphi(r)$, and $t = \psi(r)$. The solid curve $t = \psi(r)$ seems almost horizontal in (**a**), but a close look in (**b**) reveals that its intersection with the dashed line $t = \varphi(r)$ is above the dotted line $t = \alpha(r)$. Thus, $r_F < r_S$ results

anism selected. In fact, the computed values $r_F = r_S = 0.1107$ are approximately equal to $c = 0.11$.

Next, we compare social welfare. The computed values of social welfare are \$323,250 for FITs for all PV electricity and \$321,210 for FITs for surplus PV electricity in Fig. 4.4; and the values are \$283,940 for FITs for all PV electricity and \$280,300 for FITs for surplus PV electricity in Fig. 4.5. We can confirm the theoretical results: social welfare with FITs for all PV electricity is greater than that with FITs for surplus PV electricity; this is true for both of the cases depicted in Figs. 4.4 and 4.5.

4.6 Effects of Reduced Electricity Consumption

Until this point, it has been assumed that electricity consumption is not affected if any type of FIT is implemented. However, in reality, it is more likely that if either FITs for surplus PV electricity or net metering is implemented, an adopter can sell more PV electricity by decreasing self-consumption of PV electricity. In other words, these two mechanisms incentivize adopters to reduce their electricity consumption. It will thus make sense to assume that adopters reduce their electricity consumption to some extent under FITs for surplus PV electricity and net metering.

In this section, we will adapt the models developed in Sect. 4.3 and examine the effect of reduced electricity consumption on social welfare and surcharged electricity rates.

4.6.1 Definition of New Variables

Strictly speaking, the amount by which electricity consumption is reduced under FITs for surplus PV electricity and net metering may depend on the PV electricity price. However, in reality, the amount of reduction may be considered constant because most people are not so rational as to continually adjust their consumption according to the PV electricity price. Specifically, for household i, y_i decreases to $\tilde{y}_i$, while x_i increases to $\tilde{x}_i$, keeping $\tilde{x}_i + \tilde{y}_i = z_i$. Similarly, $q_{g,i}$ and $q_{c,i}$ decrease to $\tilde{q}_{g,i}$ and $\tilde{q}_{c,i}$, respectively. Accordingly, q_i decreases to $\tilde{q}_i$ so that $\tilde{q}_{g,i} + \tilde{q}_{c,i} = \tilde{q}_i$.

The loss of utility resulting from the reduction in electricity consumption may be ignored: households would not reduce electricity consumption to the degree that the losses would be significant.

4.6.2 Adapted Models

The changes to the model developed in Sect. 4.3 are systematic: the relevant parameters should be substituted with the new variables defined in Sect. 4.6.1.

First, examine FITs for surplus PV electricity. An adopter i sells $\tilde{x}_i$ at p and purchases $\tilde{q}_i - \tilde{y}_i$ at r. In other words, i purchases $\tilde{q}_i$ at r and sells $\tilde{x}_i$ at p and $\tilde{y}_i$ at r. Hence, the cost of a PV system decreases by $p\tilde{x}_i + r\tilde{y}_i$. Thus, i would adopt if and only if $v_i \geq I_i - p\tilde{x}_i - r\tilde{y}_i$. The threshold price of PV electricity $p_{S,i}(r)$ is defined as $(I_i - v_i - r\tilde{y}_i)/\tilde{x}_i$.

Net metering may be similarly examined. Adopter i sells $z_{g,i} - \tilde{q}_{g,i}$ at p during net generation periods and purchases $\tilde{q}_{c,i} - z_{c,i}$ at r during net consumption periods. In other words, i sells $z_{g,i} - \tilde{q}_{g,i}$ at p and $z_{c,i} + \tilde{q}_{g,i}$ at r, and purchases $\tilde{q}_i$ at r. Hence, the cost of a PV system decreases by $p(z_{g,i} - \tilde{q}_{g,i}) + r(z_{c,i} + \tilde{q}_{g,i})$. Thus, i would adopt if and only if $v_i \geq I_i - p(z_{g,i} - \tilde{q}_{g,i}) - r(z_{c,i} + \tilde{q}_{g,i})$. The threshold price of PV electricity $p_{M,i}(r)$ is defined as $\left[I_i - v_i - r(z_{c,i} + \tilde{q}_{g,i})\right]/(z_{c,i} - \tilde{q}_{g,i})$.

4.6.3 Surcharged Electricity Rates Revisited

To conduct an investigation that allows us to obtain an intuitive result, let us assume that the set of adopters does not change with the reduction in electricity consumption.

On the one hand, the curve $t = \alpha(r)$ changes to $t = (r-c)\left(\sum_{i\in N\setminus A} q_i + \sum_{i\in A} \tilde{q}_i\right)$ in the rt–plane (Fig. 4.3). It still passes through $(c, 0)$ but has a gentler slope. On the other hand, $t = \varphi(r)$ does not change because it corresponds to FITs for all PV electricity. Hence, r_F, which is the r-coordinate of the intersection of the two lines, will become large.

Next, consider $t = \psi(r)$, which corresponds to FITs for surplus PV electricity. It changes to $t = \sum_{i\in S(r)} (p_S(r)\tilde{x}_i + r\tilde{y}_i - cz_i)$ with the reduction in electricity consumption. Recall that y_i decreases to $\tilde{y}_i$, while x_i increases to $\tilde{x}_i$, keeping $\tilde{x}_i + \tilde{y}_i = z_i$. Because $p_S(r) > r$, this change makes $\psi(r)$ larger. In other words, curve $t = \psi(r)$ will shift upwards. Then, it is not clear how the intersection of $t = (r - c)\left(\sum_{i\in N\setminus A} q_i + \sum_{i\in A} \tilde{q}_i\right)$ and $t = \sum_{i\in S(r)} (p_S(r)\tilde{x}_i + r\tilde{y}_i - cz_i)$ moves. In other words, it is unclear whether r_S, which is the r-coordinate of the intersection, will become large or small. The same result holds if net metering is considered instead of FITs for surplus PV electricity.

4.6.4 Social Welfare Maximization Revisited

Again, we assume that the set of adopters does not change with the reduction in electricity consumption. Then, the social welfare expressed as $\sum_{i\in A} [(b + c)z_i + v_i - I_i]$, where A is F or S, does not change because it is independent of the amount of electricity consumption.

It should be noticed that while the amount of PV generation $\sum_{i\in A} z_i$ does not change, electricity consumption decreases by $\sum_{i\in A} (q_i - \tilde{q}_i)$. The decrease in electricity consumption may reduce the conventional generation costs and reduce greenhouse gas emissions. Hence, the reduction in electricity consumption may increase social welfare in this regard, although this effect does not appear in $\sum_{i\in A} [(b + c)z_i + v_i - I_i]$.

This increase in social welfare becomes large as the reduction $\sum_{i\in A} (q_i - \tilde{q}_i)$ increases. Hence, the result in Sect. 4.4.2 should be modified in this case. If the reduction in electricity consumption is not considered, the greatest social welfare is achieved by FITs for all PV electricity. However, if the reduction is taken into account and if the reduction is sufficiently large, the greatest social welfare may be achieved by FITs for surplus electricity or net metering, rather than by FITs for all PV electricity.

4.7 Conclusion

This chapter examined differences among FITs for all PV electricity, FITs for surplus PV electricity, and net metering. We developed a simple microeconomic model, characterized each mechanism, calculated the price of PV electricity, and compared the mechanisms with respect to surcharged electricity rates and social welfare. A numerical example was also provided to illustrate the theoretical results.

This study presents a novel analysis not only in that previous studies have not investigated the mechanisms comparatively but also in that they have predominantly offered empirical analyses or qualitative discussion from political perspectives.

The findings are summarized as follows. Suppose that electricity consumption does not change regardless of which type of mechanism is implemented. First, it is unclear which mechanism yields the lowest electricity rate if the electric utility's cost of purchasing PV electricity is passed on to all ratepayers by increasing the retail electricity rate. However, if households are more homogeneous in terms of relevant parameters, the surcharged electricity rate is almost the same regardless of the mechanism selected. Second, FITs for all PV electricity produce the greatest social welfare.

Next, suppose that electricity consumption will be reduced to some extent if FITs for surplus PV electricity or net metering is implemented. Then, the first result is not affected: which mechanism yields the lowest surcharged electricity rate is unclear. However, the second result may require modification. Which mechanism produces the largest social welfare depends on the amount by which electricity consumption is reduced. On the one hand, if the reduction is relatively small, FITs for all PV electricity are likely to produce the largest social welfare. On the other hand, if the reduction is significantly large, FITs for surplus PV electricity or net metering may produce larger social welfare than FITs for all PV electricity.

As described above, the results, particularly the surcharged electricity rates, depend on the parameter values of households. In our model, these include the amount of PV electricity as well as surplus and self-consumption of PV electricity, the satisfaction a household experiences from the adoption, and the cost of a PV system. Accordingly, numerical examination would play a role in providing a more clear-cut solution to this problem. In fact, our numerical example indicates there may be few differences in the surcharged electricity rate. In this regard, empirical investigation based on a real set of parameter values might be necessary, particularly to investigate the effect of reduced electricity consumption when a certain mechanism is implemented.

The study presented in this chapter may be considered as a starting point for proposing a new way of pricing RES-E. Particularly when a new technology becomes available, a new pricing method would be much more important because the conventional system may no longer be effective. For example, a smart grid system has been under development; this is one of the power distribution systems of the future that will employ information and communications technology to optimize electricity supply and demand. Such a system will require new pricing mechanisms, thus presenting new theoretical challenges.

Appendix 1: Proof of $p_S(p_F) = p_F$ and $S(p_F) = F$

To make the description as simple as possible, assume $p_{F,i} < p_{F,i+1}$ for any i by, if necessary, renaming each household. In particular, $p_F = p_{F,n}$. Because $p_{S,i}(p_{F,i}) = p_{F,i}$ for any i from Eqs. (4.1) and (4.2) as well as $x_i + y_i = z_i$, $p_{S,n}(p_F) = p_F$ follows.

Then, $p_{S,h}(p_F) < p_{S,n}(p_F) < p_{S,k}(p_F)$ holds for any $h < n$ and any $k > n$: recall that for any i, $p_{S,i}(r)$ is a decreasing function of r. The first inequality, $p_{S,n}(p_F) > p_{S,h}(p_F)$, is yielded from $p_{S,n}(p_F) = p_F$, $p_F > p_{F,h}$, $p_{F,h} = p_{S,h}(p_{F,h})$, and $p_{S,h}(p_{F,h}) > p_{S,h}(p_F)$. The second inequality is similarly shown.

Hence, $p_{S,n}(p_F)$ is the nth smallest $p_{S,i}(p_F)$ of $\{p_{S,i}(p_F)|i \in N\}$, that is, $p_{S,n}(p_F) = p_S(p_F)$. Because $p_{S,n}(p_F) = p_F$, $p_S(p_F) = p_F$ holds. Then, the condition for household i to adopt a PV system under FITs for surplus PV electricity, i.e., $v_i \geq I_i - px_i - ry_i$, becomes $v_i \geq I_i - p_F z_i$. This is exactly the condition for i to adopt a PV system under FITs for all PV electricity. Hence, $S(p_F) = F$ holds.

Similarly, we can prove $p_M(p_F) = p_F$ and $M(p_F) = F$.

Appendix 2: An Alternative Mechanism

Let all PV electricity be purchased at $b + c$. Based on this pricing, if exactly n households adopt PV systems, we would be done. Otherwise, either fewer or more than n households will adopt. On the one hand, if fewer than n households adopt, the electric utility is required to pay a fixed amount to adopters, such that exactly n households adopt. On the other hand, if more than n households adopt, adopters must pay a fixed fee to the electric utility so that exactly n households adopt.

This mechanism can easily be shown to produce more social welfare than FITs for all PV electricity. First, consider the case in which the electric utility must pay a fixed payment. Let T be this fixed payment. Define $T_i = I_i - v_i - (b + c)z_i$. Then, T may be set equal to the nth smallest T_i so that exactly n households adopt PV systems. Note that for those n households, $(b + c)z_i + v_i - I_i + T \geq 0$.

Next, compare social welfare between this mechanism and FITs for all PV electricity. Define $G = \{i \in N | T_i \leq T\}$, which is the set of n adopters under this mechanism. Note that if and only if $i \in G$, $-T_i = (b+c)z_i + v_i - I_i \geq -T$ holds. Thus, social welfare, $\sum_{i \in G} [(b + c)z_i + v_i - I_i]$, is obviously no less than $\sum_{i \in F} [(b + c)z_i + v_i - I_i]$, the social welfare under FITs for all PV electricity.

In the case in which adopters must pay a fixed fee, household i's threshold fee is defined as $T_i = (b + c)z_i + v_i - I_i$. Set the fee T at the nth largest T_i. The proof proceeds in the same way.

The idea underlying this alternative mechanism is applied in Chap. 5.

References

Black AJ (2004) Financial payback on California residential solar electric systems. Sol Energy 77(4):381–388

Bürer MJ, Wüstenhagen R (2009) Which renewable energy policy is a venture capitalist's best friend? Empirical evidence from a survey of international cleantech investors. Energy Policy 37(12):4997–5006

Carley S (2009) Distributed generation: an empirical analysis of primary motivators. Energy Policy 37(5):1648–1659

Couture T, Gagnon Y (2010) An analysis of feed-in tariff remuneration models: implications for renewable energy investment. Energy Policy 38(2):955–965

Duke R, Williams RH, Payne A (2005) Accelerating residential PV expansion: demand analysis for competitive electricity market. Energy Policy 33(15):1912–1929

EIA (2017) Sources & users. www.eia.gov/electricity/sales_revenue_price/pdf/table6.pdf. Accessed 26 Feb 2017

Fouquet D, Johansson TB (2008) European renewable energy policy at crossroads: focus on electricity support mechanisms. Energy Policy 36(11):4079–4092

Hughes L, Bell J (2006) Compensating customer-generators: a taxonomy describing methods of compensating customer-generators for electricity supplied to the grid. Energy Policy 34(13):1532–1539

IEA (2012) World energy outlook 2012. IEA, Paris

JPEA (2015) A leaflet of residential PV generation. www.jpea.gr.jp/pdf/susume_pamphlet.pdf. Accessed 26 Feb 2017 (In Japanese)

Menanteau P, Finon D, Lamy ML (2003) Prices versus quantities: choosing policies for promoting the development of renewable energy. Energy Policy 31(8):799–812

Mendonça M (2007) Feed-in tariffs: accelerating the deployment of renewable energy. Earthscan, London

METI (2013) The report on the diffusion of PV systems in Japan. www.meti.go.jp/meti_lib/report/2013fy/E002502.pdf. Accessed 6 July 2017 (In Japanese)

Mills A, Wiser R, Barbose G, Golove W (2008) The impact of retail rate structures on the economics of commercial photovoltaic systems in California. Energy Policy 36(9):3266–3277

Mitchell C, Bauknecht D, Connor PM (2006) Effectiveness through risk reduction: a comparison of the renewable obligation in England and Wales and the feed-in system in Germany. Energ Policy 34(3):297–305

NEDO (2017) MONSOLA-11. app0.infoc.nedo.go.jp/metpv/monsola.html. Accessed 7 July 2017 (In Japanese)

Chapter 5
Feed-in Tariffs Combined with Capital Subsidies

Abstract Both feed-in tariffs (FITs) and capital subsidies have been widely employed to promote the adoption of renewable energy technologies. This chapter sheds light on the combined use of FITs and capital subsidies. The purpose is to clarify their optimal combinations to encourage households to adopt photovoltaic (PV) systems or to encourage firms to invest in PV generation. This study develops a microeconomic model embodying the idea of two-part tariffs. The most important findings concern the combination that maximizes social welfare for the residential sector: if FITs are applied to the total PV electricity generated, they should be set at the avoided cost per unit of PV electricity, and capital subsidies should be used to control the number of adopters; whereas, if FITs are applied to only surplus PV electricity, the previous principle is distorted to some extent. A similar result is obtained for the business sector. In the model for the business sector, the government aims to have a certain installed capacity of PV panels, whereas in the model for the residential sector, its aim is to have a certain number of households adopt PV systems. The problem of equity, that is, how to finance the cost of FITs and capital subsidies is also discussed.

Keywords Feed-in tariff · Capital subsidy · Two-part tariff · Photovoltaic system

5.1 Introduction

A wide variety of support measures have been employed to promote renewable energy technologies (RETs), including, among others, renewable portfolio standards, tax credits, and the subjects of this chapter: feed-in tariffs (FITs) and investment or capital subsidies. A typical FIT system allows the electricity generated from renewable energy sources (RES-E) to be sold to electric utilities at a set price for a set period

---

Modified, with permission of Elesevier, from Yamamoto, Y., Feed-in tariffs combined with capital subsidies for promoting the adoption of residential photovoltaic systems, Energy Policy, 111, 312–320, Elsevier, 2017. I would like to thank Elsevier.

Y. Yamamoto, *Feed-in Tariffs and the Economics of Renewable Energy*,
https://doi.org/10.1007/978-3-319-76864-9_5

of years (Mir-Artigues and del Río 2016, Chap. 8); both the price and period are determined by a regulatory authority, and the cost of purchasing the RES-E is passed on to all ratepayers through increased electricity rates. In contrast, capital subsidies are granted to households or businesses as a lump sum payment when they adopt RETs; the funds for capital subsidies are raised through taxation. These systems serve as incentives for adopting RETs and generating RES-E, with the goal of promoting diffusion of the technologies.

FITs and capital subsidies are often used in combination (Dusonchet and Telaretti 2010; Hsu 2011; Zhao et al. 2013; del Río and Mir-Artigues 2014; Mir-Artigues and del Río 2014). A possible reason for this is that each offers distinct advantages (del Río and Mir-Artigues 2014; Mir-Artigues and del Río 2014; Hirvonen et al. 2015). FITs ensure relatively stable revenue flows to investors over a predetermined period of years; moreover, they enable a government to spread its expenditure more evenly over that period, which may enhance public acceptance, resulting in the political feasibility of this approach. In contrast, capital subsidies can lower an investor's cost of financing, which is the main barrier to investment.

However, the combination of FITs and capital subsidies may involve either inefficiency, redundancy, or overlap (del Río and Mir-Artigues 2014; Mir-Artigues and del Río 2014). If FITs and capital subsidies are not coordinated properly, the combination may either overcompensate adopters, thereby burdening ratepayers and taxpayers, or undercompensate, thereby potentially failing to achieve the target degree of RET diffusion (Lesser and Su 2008; Mayr et al. 2014).

In Japan, for example, the number of households that have adopted photovoltaic (PV) systems has increased enormously since 2009, when both FITs and capital subsidies became widely available (METI 2013). In addition to the national government, many prefectural and municipal governments have offered households capital subsidies. In 2013, the Japanese government provided a capital subsidy of either ¥15,000/kW or ¥20,000/kW per installed capacity of PV panels (¥100.00 was approximately equivalent to $1.00 at the time). Those adopters whose cost of a PV system was less than or equal to ¥500,000/kW were granted ¥15,000/kW, and those whose cost was less than or equal to ¥410,000/kW were granted ¥20,000/kW. To encourage cost reductions, the subsidy was larger for those whose cost was lower. In addition, the Gunma prefectural government and the city of Takasaki, Gunma, for instance, offered capital subsidies of ¥15,000/kW and ¥30,000/kW, respectively. Furthermore, households could sell PV electricity at much higher FITs than was the case with retail electricity rates: PV electricity was priced at ¥38.0/kWh, while retail electricity rates were either ¥18.89, ¥25.19 or ¥29.10/kWh, depending on the amount of monthly consumption (Tokyo Electric Power Company, June 2013). Note that monthly electricity consumption was divided by 120 and 300 kWh into three tiers; a higher rate was applied to the higher tier of consumption. For example, if monthly consumption was 400 kWh, the bill was the sum of ¥18.89/kWh multiplied by 120 kWh, ¥25.19/kW multiplied by 180 kWh, and ¥29.10/kW multiplied by 100 kWh. It seems that these compensations were so generous that PV systems were adopted even if physical conditions, e.g., solar radiation, were unsuitable, placing an excessive financial burden on society.

Nevertheless, only a limited number of studies (reviewed in Sect. 5.2.1) have considered FITs and capital subsidies in combination, relative to the voluminous literature investigating such support measures in isolation. The purpose of this chapter is to address this gap in the existing research, in particular by developing a theoretical model.

This study focuses on FITs and capital subsidies mainly for the residential sector, where households considering adoption of RETs, typically PV systems, are interested in such remunerations, but, at the same time, must pay for FITs and capital subsidies indirectly through increased electricity rates and taxes. Taking into account that these support measures are aimed at diffusing PV systems further, the study focuses on the earlier stage of diffusion by assuming that the amount of PV generation is much smaller than the electricity consumption of the population.

This study develops a microeconomic model, which focuses on decision-making regarding the adoption of a PV system. It aggregates engineering details with a simple parameter whenever possible (cf. Mir-Artigues and del Río 2016, Chap. 7) The present model is as an extension of the model by Yamamoto (2012); it incorporates both FITs and capital subsidies to investigate them in combination, and it may be outlined as follows. A government offers a combination of FITs and capital subsidies to households for the adoption of a PV system. Then, households make decisions about adoption based on utility maximization. The model is not dynamic but rather static, as its aim is to examine decision-making by heterogeneous households at present, given the availability of FITs and capital subsidies.

The model assesses the relative superiority of different combinations of FITs and capital subsidies against several common governmental criteria (Mir-Artigues and del Río 2016, Chap. 5). These criteria include maximization of PV electricity output in terms of the benefit it provides (Oliva et al. 2014), minimization of promotion costs in terms of cost-effectiveness (del Río and Cerdá 2014; Cerdá and del Río 2015), and maximization of social welfare (e.g., Pirnia et al. 2011). In particular, social welfare is commonly used in the economics literature to assess economic efficiency; it consists of the surplus or net benefit to market participants, typically consumers and producers, in equilibrium. Pirnia et al. (2011) provided a simple explanation for social welfare on the electricity market. A concrete definition of social welfare is presented in Sect. 5.4.3 of this study.

The remainder of the chapter is organized as follows. Section 5.2 reviews the relevant literature, including the idea of a two-part tariff, which is applicable to the combination of FITs and capital subsidies. Section 5.3 describes a basic model, where FITs are applied to all of the PV electricity generated. Section 5.4 examines the model in terms of several governmental criteria. Section 5.5, in contrast, addresses FITs applied only to surplus PV electricity by adapting the model developed in Sect. 5.3, and the section revisits the social welfare maximization problem. In contrast, in Sect. 5.6, we briefly consider a model for the business sector. Section 5.7 offers a discussion of the results in terms of the previous research. Finally, Sect. 5.8 provides the policy implications of the results and concludes with suggestions for future research.

5.2 Literature Review

This section offers a review of the relevant literature. In the first subsection, studies dealing with the combined use of FITs and capital subsidies are reviewed. Unlike those studies, the present study views the problem as a variation on a two-part tariff system. The second subsection reviews literature relevant to the idea of two-part tariffs and identifies several hurdles that must be overcome if the idea is to be applied to our problem.

5.2.1 Studies on the Combined Use of FITs and Capital Subsidies

To my knowledge, only two studies have explicitly addressed the combined use of FITs and capital subsidies, although a few others address the issue implicitly. The first of the two explicit studies is by Mir-Artigues and del Río (2014), who developed a dynamic mathematical model to account for the inter-temporality of FITs and upfront capital subsidy payments. They found that, for an investor to achieve a given level of profitability, i.e., a ratio of profit to investment, financial costs are invariable for any combination of FITs and capital subsidies. What should be investigated in the next step is the financial cost to all investors; in general, given a certain combination of FITs and capital subsidies, some potential investors invest and some do not, depending on their own profitability calculations.

The second study, by Hsu (2011), examined the combination through a system dynamics model that simulates both the adoption of PV systems in society and the policy cost. Hsu compared three combinations: high FITs and low subsidies, medium FITs and medium subsidies, and low FITs and high subsidies. It was found that there is no difference among the three in terms of adoption of PV systems but that a lower policy cost is achieved with the lower FITs/higher capital subsidies combination. The causal feedback in the system dynamics model led to the results, but each household's decision-making was not addressed explicitly in that study.

Two studies have dealt with the issue implicitly. Both of these sought to make use of an auction process. First, Mayr et al. (2014) investigated an auction for qualifying eligible developers and determining remuneration levels for each support scheme. Second, Lesser and Su (2008) proposed a two-part FIT scheme consisting of a capacity payment and an energy payment. The capacity payment level is determined through an auction process, whereas energy payments are tied to the spot market price of electricity. As these studies show, an auction allows policymakers to avoid having to determine FIT or capital subsidy levels administratively. However, the scheme would be somewhat complicated if an auction were applied to the combination of FITs and capital subsidies.

Finally, it deserves mentioning that Klein (2008) described a methodology to determine FIT levels in terms of two concepts of cost associated with RES-E. One is the cost of RES-E generation, including investments in production facilities, O&M costs, and interest payments for loans. The other is the cost avoided by using RES-E, including not only the expenses of conventional power generation but also external costs—costs not priced in through market mechanisms—such as those associated with climate change, health impacts of air pollution, and energy supply vulnerability. These concepts of cost clarify two distinct objectives of compensation: RES-E generation cost compensation and avoided cost compensation. However, governments set only a single FIT level, resulting in the present controversy about FIT levels. One avenue to reconciliation may be to use two types of remuneration, one for RES-E generation-cost compensation, and the other for avoided-cost compensation. In other words, it is suggested that FITs might be used for avoided-cost compensation and that capital subsidies might be used for RES-E generation-cost compensation.

In light of the above literature review, the idea of two-part tariffs may play a key role in appropriately setting the respective levels of FITs and capital subsidies. Therefore, Sect. 5.2.2 reviews the standard understanding of two-part tariffs.

5.2.2 *Two-Part Tariffs*

Oi (1971) offered a seminal investigation of two-part tariffs that makes the concept clear. Suppose that a monopoly producer sets the price of a good or a service to maximize its profit. If the good or service cannot be resold, as in the case of, for example, an amusement park ride, the producer can enjoy greater profits by using a two-part tariff than would be the case otherwise. A two-part tariff consists of a use price (the price charged per ride at an amusement park, for example) and a lump sum charge (an admission fee for the right to purchase rides at the use price). It is prescribed to the producer that the use price should be used to achieve profit maximization or economic efficiency, whereas the lump sum charge should be used to attain a target amount of market participation.

Now, let us consider applying the idea of two-part tariffs to our problem of setting optimal levels of FITs and capital subsidies. On the one hand, FITs, paid for the RES-E generated, may be regarded as corresponding to a use price in the two-part tariff model. On the other hand, capital subsidies, paid up front for the installation of a PV system, may be regarded as corresponding to a lump sum charge in that model.

However, in this approach, it must be acknowledged that households are incapable of choosing the amount of PV electricity they will generate if they adopt PV systems. In other words, when a household is offered FITs and capital subsidies, it only decides whether to adopt a PV system. In addition, the sets of adopters may vary, depending on the combination of FITs and capital subsidies. Thus, the first-order condition for optimization is no longer applicable to our situation. Considering these distinctive features of our problem, we develop a model in Sect. 5.3.

Table 5.1 Notations

p	Price of PV electricity
s	Capital subsidy
I_i	Household i's cost of a PV system
v_i	Household i's satisfaction obtained from adopting a PV system
z_i	Household i's amount of PV generation
x_i	Household i's amount of surplus PV electricity
y_i	Household i's amount of self-consumed PV electricity
m_i	Site i's PV capacity in the model for the business sector
b	Benefit of the avoided external cost per unit of PV electricity
c	Unit cost of conventional power generation
n	Targeted number of adopters
Ω	Set of adopters if FITs are applied to all PV electricity
Θ	Set of adopters if FITs are applied to surplus PV electricity
r	Surcharged electricity rate
L	Electricity consumption by all households

5.3 Basic Model

The model to be developed in this section assumes that FITs are applied to all of the PV electricity generated. This assumption can simplify the analysis and serve as a basis for the investigation in Sect. 5.5, where FITs are only applied to surplus PV electricity, i.e., the PV electricity that is not self-consumed and is fed into the power grid.

5.3.1 Definition of Variables

This subsection provides definitions of the variables that appear in the model, which are essentially the same as those for the residential sector in Chaps. 3 and 4. We describe them again below, together with some modifications added to some of them in this chapter. Key notations are summarized in Table 5.1.

A government provides two incentives, FITs and capital subsidies. If a household adopts a PV system, it can sell all the PV electricity it generates to an electric utility at price p for a given period of years. For simplicity, it is presumed that the period for FITs equals the lifetime of a PV system. Furthermore, a household is eligible for capital subsidy s when purchasing and installing the system. Both p and s are determined by the government.

It is plausible that there may be few differences in the installed capacity of PV panels, m_i, among adopters due to their having similar areas on their rooftops available for the installation of a PV system. Thus, we express the values of a PV system cost $m_i I_i$, an amount of PV electricity generated $m_i z_i$, and satisfaction perceived $m_i v_i$ as I_i, z_i, and v_i, respectively. Put differently, it is defined in this chapter that for a household i, the cost of a PV system is I_i, the amount of PV electricity generated is z_i, and the satisfaction i experiences from the adoption is v_i.

A discount rate is taken into account for z_i so that household i can obtain the payment pz_i for the sale of PV electricity throughout the period (see Appendix 1 in Chap. 3). It is assumed that $I_i > v_i$, which, as will be clear in Sect. 4.3.2, implies that no households adopt PV systems without either FITs or capital subsidies.

Let b and c represent the benefit of the avoided external costs (realized per unit of PV electricity) and the unit cost of conventional power generation, respectively.

Finally, it is assumed that a government aims at incentivizing—as efficiently as possible—exactly n households to adopt PV systems by means of FITs and capital subsidies.

5.3.2 *Household Decision-Making*

A household i does not adopt a PV system unless the investment in PV generation pays. In other words, it will adopt if and only if p and s are offered such that $pz_i + v_i + s \geq I_i$ (see Appendix 2 in Chap. 3). A downward-sloping line

$$s = -pz_i - v_i + I_i \quad (p \geq 0) \tag{5.1}$$

forms a boundary in the ps–plane between i's adopting and not adopting. Let $\Omega_{p,s}$ denote the set of households that adopt if (p, s) is offered, i.e., $\Omega_{p,s} = \{i \mid pz_i + v_i + s - I_i \geq 0\}$. Note that the tax and the surcharge on the retail electricity rate do not appear in Eq. (5.1) because every household must incur these burdens regardless of whether it adopts a PV system.

5.3.3 *Potential Combinations of FITs and Capital Subsidies*

The locus of (p, s), under which exactly n households adopt PV systems, may be identified using the following procedure (Fig. 5.1). First, draw line (5.1) for every household i in the ps–plane. Second, find the nth smallest s–coordinate $I_i - v_i$ on the s–axis, Point A in Fig. 5.1. Third, starting from this point, trace the corresponding line (5.1) toward the lower right until it intersects any other line. Fourth, transfer to the other line at that intersection, Point B in Fig. 5.1, and start tracing the latter line toward the lower right. Fifth, continue the same procedure beyond the p–axis, Point Z in Fig. 5.1. This procedure yields a broken line, $s = f(p)$. More than n households will adopt at the bending points of $s = f(p)$, such as C in Fig. 5.1, where at least two lines corresponding to different households intersect. For convenience, it is assumed

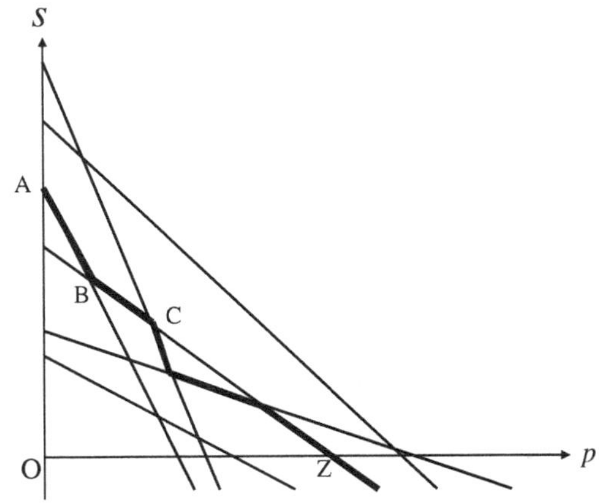

Fig. 5.1 The locus of (p, s). The broken bold line is the locus of (p, s), under which exactly n households adopt PV systems, shown for the case $n = 4$ for illustration. The locus forms the curve $s = f(p)$, the level of subsidy for a given price of PV electricity to have $n = 4$ households adopt PV systems

that one household is chosen based on a rule, for instance, first-come, first-served. Negative values of $f(p)$ after Point Z mean that lump sum charges are incurred for the right to feed PV electricity into the grid, rather than subsidies being offered.

The government chooses (p, s), only satisfying $s = f(p)$ in view of efficiency, although exactly n households may adopt even if p or s is slightly increased. The household that is the first to no longer adopt, as either p or s decreases infinitesimally, is denoted as the *marginal household*. Let Ω_p denote $\Omega_{p,f(p)}$.

5.4 Optimal Combinations Based on Each Criterion

Let us calculate optimal combinations of FITs and capital subsidies by each of the criteria: maximization of PV electricity, minimization of promotion cost, and maximization of social welfare.

5.4.1 *Maximization of PV Electricity*

Suppose the government seeks to maximize the amount of PV electricity n adopters will generate. As the amount of PV electricity to be generated is $\sum_{i \in \Omega_p} z_i$, the PV electricity maximization problem is

$$\text{maximize} \sum_{i \in \Omega_p} z_i. \tag{5.2}$$

The government should set a higher p because the amount of PV electricity becomes large as p is set higher, as shown in the following. Suppose (p, s) is traveling

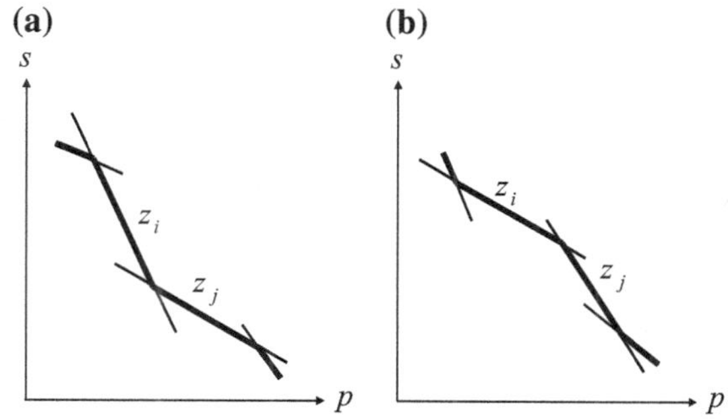

Fig. 5.2 Two possibilities of (p, s) traveling on $s = f(p)$. The curve $s = f(p)$ has two possibilities with respect to a link: **a** $z_i > z_j$ and **b** $z_i < z_j$. If (p, s) is traveling from a line segment of i to the adjacent line segment of j, **a** both households i and j still adopt PV systems; or **b** household j adopts a system, while household i no longer adopts

along $s = f(p)$ from Point A in Fig. 5.1 towards the lower right. As long as (p, s) remains on any one of the same line segments of $s = f(p)$, $\sum_{i \in \Omega_p} z_i$ is constant because the members in Ω_p are unchanged.

Now, suppose (p, s) is traveling from one line segment to an adjacent segment. There are two possibilities. The first is that the slope z_i of the anterior line segment is larger than the slope z_j of the posterior line segment, i.e., $z_i > z_j$ (Fig. 5.2a), as at Point B in Fig. 5.1. In this case, only the marginal household changes from i to j, while the set Ω_p of adopters does not change. Hence, $\sum_{i \in \Omega_p} z_i$ is invariable. The second possibility is the reverse, i.e., $z_i < z_j$ (Fig. 5.2b), as at Point C in Fig. 5.1. In this case, Ω_p does change. On the one hand, household j, which did not adopt for the (p, s) on the anterior line segment, comes to adopt and becomes the marginal household for the (p, s) on the posterior line segment. On the other hand, household i, the marginal household on the anterior line segment, no longer adopts for the (p, s) on the posterior line segment. Because $z_i < z_j$, $\sum_{i \in \Omega_p} z_i$ increases through this process. Taken together, $\sum_{i \in \Omega_p} z_i$ never decreases, while p increases from zero.

If the government that aims to maximize the amount of PV electricity does not intend to make households incur a lump sum charge, it should choose the nth smallest $p = (I - v_i)/z_i$ together with $s = 0$, Point Z in Fig. 5.1.

5.4.2 *Minimization of Promotion Cost*

Suppose the government seeks to minimize the promotion cost, i.e., the cost of encouraging households to adopt PV systems. As the promotion cost consists of the cost of FITs, $p \sum_{i \in \Omega_p} z_i$; the cost of capital subsidies, $nf(p)$; minus the avoided cost of conventional power generation, $c\sum_{i\in\Omega_p} z_i$, the promotion cost minimization problem is

$$\text{minimize } (p-c)\sum_{i\in\Omega_p} z_i + nf(p). \tag{5.3}$$

It is not clear whether the promotion cost increases, decreases, or remains constant when p increases, as shown in the following. Suppose (p, s) is currently on one of the line segments of $s = f(p)$, the slope of which is z_j. Differentiate $(p-c)\sum_{i\in\Omega_p} z_i + nf(p)$ with respect to p, using $f'(p) - z_j$. The result, $-nz_j + \sum_{i\in\Omega_p} z_i$, has a fixed sign because the set Ω_p of adopters does not change as long as (p, s) is on the same line segment. However, the sign itself depends on which is greater, $\sum_{i\in\Omega_p} z_i/n$ or z_j. This is indefinite because $\sum_{i\in\Omega_p} z_i/n$ is the mean, $\bar{z}$, of z_i's over Ω_p, one member of which has z_j. Hence, it is not clear whether the promotion cost is increasing, decreasing, or invariable when (p, s) is traveling along a line segment towards the lower right.

Next, suppose (p, s) is traveling from one line segment to an adjacent segment. In the case depicted by Fig. 5.2a, the mean $\bar{z}$ does not change, while z_i decreases to z_j. Accordingly, if $\bar{z} - z_i > 0$, then $\bar{z} - z_j > 0$. However, if $\bar{z} - z_i < 0$, it may be $\bar{z} - z_j > 0$ or $\bar{z} - z_j < 0$. Hence, the sign is not definite. In the case depicted in Fig. 5.2b, while z_i increases to z_j, $\bar{z}$ increases from $\bar{z}_{(i)}$ to $\bar{z}_{(j)}$ by $\left(z_j - z_i\right)/n$, where $\bar{z}_{(i)}$ represents the mean $\bar{z}$ over Ω_p that contains household i. Accordingly, if $\bar{z}_{(i)} - z_i < 0$, then $\bar{z}_{(j)} - z_j < 0$. However, if $\bar{z}_{(i)} - z_i > 0$, it may be that $\bar{z}_{(j)} - z_j > 0$ or $\bar{z}_{(j)} - z_j < 0$. Hence, the sign is also not definite in this case. The promotion cost, therefore, depends on the parameter values.

Note that the same result follows if the promotion cost is defined as being exclusive of the conventional generation cost savings, i.e., $p\sum_{i\in\Omega} z_i + nf(p)$.

5.4.3 *Maximization of Social Welfare*

Suppose the government seeks to maximize social welfare, which is defined as the sum of consumer surplus for all households, producer surplus for an electric utility, net revenue for a government, and societal benefit. Let social welfare be normalized to zero when neither FITs nor capital subsidies are offered.

The consumer surplus of a household is the sum of FIT payments for PV electricity generated and capital subsidies, minus the cost of a PV system, the electricity surcharges covering the FIT cost, and the tax covering the capital subsidization cost. The producer surplus is the revenue from the surcharges and the avoided costs of conventional power generation minus the cost of FIT payments. The government's net revenue is the tax revenue minus the capital subsidies to the adopters. Finally, the societal benefit is realized by using PV electricity, as defined in Sect. 5.3.1.

Note that the following terms cancel each other out: the electricity surcharges between households and the electric utility, the FIT payments between the electric utility and the adopters, the capital subsidies between the government and the adopters, and the tax payments between households and the government.

As a result, the social welfare amounts to $\sum_{i\in\Omega_p}[(b+c)z_i - I_i]$, the sum of the societal benefit that PV electricity provides and the avoided cost of conventional power generation, less the cost of PV systems. Note that the surcharges and taxes do not appear explicitly in the expression of social welfare.

Thus, the social welfare maximization problem is

$$\text{maximize} \sum_{i\in\Omega_p}[(b+c)z_i - I_i]. \tag{5.4}$$

The solution to (5.4) is $p = b + c$, which yields $s = f(b + c)$. This is verified as follows. Social welfare $\sum_{i\in\Omega_p}[(b+c)z_i - I_i]$ changes if and only if the set Ω_p of adopters changes, as depicted in Fig. 5.2b. Suppose that when p increases to p_1, the marginal household changes from i to j. Plugging $p = b+c$ into Eq. (5.1) yields $-(b+c)z_i + I_i > -(b+c)z_j + I_j$, or $(b+c)z_i - I_i < (b+c)z_j - I_j$ if $p_1 < b+c$ and that $-(b+c)z_i + I_i < -(b+c)z_j + I_j$, or $(b+c)z_i - I_i > (b+c)z_j - I_j$ if $p_1 > b+c$. Because the adopters, other than the marginal households i or j, are the same, social welfare increases if $p_1 < b+c$ and decreases if $p_1 > b+c$. Therefore, it is maximized at $p = b+c$. Note that any p close to $b+c$ yields the same maximum as long as $\Omega_p = \Omega_{b+c}$.

The result indicates that for social welfare maximization, the FITs should be set at the avoided cost per unit of PV electricity, while capital subsidies should be set to control the number of adopters.

5.5 FITs Applied to Surplus PV Electricity

Up to this point, it was assumed that FITs were applied to all PV electricity. In this section, it is assumed that only surplus PV electricity is eligible for FITs. The model is adapted to allow for this change in Sect. 5.5.1, and, in view of this modification, the social welfare maximization problem is then revisited in Sect. 5.5.2.

5.5.1 Adapted Model

A household i's amount of PV electricity z_i is divided into the amount not self-consumed, namely the surplus, x_i, and the amount self-consumed, y_i, where $x_i + y_i = z_i$.

Let r denote the electricity rate, which consists of the standard electricity rate c and the surcharge to raise funds for FITs. This implies that an adopter i sells x_i at p and y_i at r, while i pays r for every unit of its electricity consumption. As will be discussed below, the government sets r, as well as p and s, as a regulator.

Given a combination of p, r, and s, a household i adopts a PV system if and only if $px_i + ry_i + v_i + s \geq I_i$. Let $\Theta_{p,r,s}$ denote the set of households that adopt if a combination of p, r, and s is offered, i.e., $\Theta_{p,r,s} = \{i \mid px_i + ry_i + v_i + s - I_i \geq 0,\ p \geq 0, r \geq 0\}$. A plane

$$px_i + ry_i + v_i + s - I_i = 0 \quad (p \geq 0,\ r \geq 0) \tag{5.5}$$

forms a boundary in the prs–space between i's adopting and not adopting.

A combination of p, r, and s must be subject to two constraints. The first constraint is that exactly n households adopt PV systems. Letting $N\left(\Theta_{p,r,s}\right)$ denote the number of members in $\Theta_{p,r,s}$, this constraint is written as

$$N\left(\Theta_{p,r,s}\right) = n. \tag{5.6}$$

From Eq. (5.6), s is yielded as a function of p and r. This process is illustrated as follows. Draw the plane (5.5) for every household in the prs–space. Then, given a pair of p and r, the nth smallest s can be found, which is denoted by $s = g(p, r)$ and corresponds to $f(p)$ in Sect. 5.3.

The second constraint is that the electric utility collects the funds for FITs through the electricity rate by adding a surcharge to the standard electricity retail rate c. Let L be the fixed sum of electricity consumption by all households (adopters plus non-adopters). As the total amount of electricity sales by the utility is $L - \sum_{i\in\Theta_{p,r,s}} y_i$, the second constraint is written as

$$(r - c)\left(L - \sum_{i\in\Theta_{p,r,s}} y_i\right) = (p - c)\sum_{i\in\Theta_{p,r,s}} x_i. \tag{5.7}$$

Note that constraint (5.7) matters in the current model because, in addition to p, r affects a household's decision-making about whether to adopt a PV system. It should be recalled that if FITs are applied to all PV electricity, as in Sects. 5.3 and 5.4, r has no effect.

Plugging in $s = g(p, r)$ reduces Eq. (5.7) to an equation of p and r. The equation may then be solved for r as $r = r(p)$, on the sensible assumption that such a function exists. Define Θ_p as $\Theta_p = \Theta_{p,r(p),g(p,r(p))}$.

5.5.2 *Maximization of Social Welfare Revisited*

Let us consider the social welfare maximization problem with the adapted model:

$$\text{maximize} \sum_{i \in \Theta_p} [(b + c)(x_i + y_i) + v_i - I_i]. \tag{5.8}$$

It will be difficult to solve (5.8) in general, as examined in the following. To begin, $r(p) \approx c$ may be deduced from Eqs. (5.6) and (5.7): because this study focuses on the earlier stage of diffusion of PV systems, L can be considered much larger than $\sum_{i \in \Theta_p} (x_i + y_i)$, so that $r(p) - c = (p - c) \sum_{i \in \Theta_p} x_i / \left(L - \sum_{i \in \Theta_p} y_i\right) \approx 0$ will follow.

Then, supposing $r(p) = c$ as an approximation, let us examine how $\sum_{i \in \Theta_{p,c,g(p,c)}} [(b + c)(x_i + y_i) + v_i - I_i]$ varies. The following lemma is helpful: on the one hand, $\sum_{i \in \Theta_{p,c,g(p,c)}} [c(x_i + y_i) + v_i - I_i]$ is maximized at $p = c$; on the other hand, how $\sum_{i \in \Theta_{p,c,g(p,c)}} b(x_i + y_i)$ changes depends on the parameter values (Appendix 1). Therefore, it will generally be impossible to decide how $\sum_{i \in \Theta_{p,c,g(p,c)}} [(b + c)(x_i + y_i) + v_i - I_i]$ varies; it depends on the parameter values. This suggests that $\sum_{i \in \Theta_p} [(b + c)(x_i + y_i) + v_i - I_i]$ varies, depending on the parameter values.

A comparison of the maximized social welfare merits investigation of the two cases: FITs applied to surplus PV electricity (this section) in relation to FITs applied to all PV electricity (Sect. 5.4.3). In general, the sufficient condition for $\sum_{i \in \Theta_{p,r,g(p,r)}} [(b + c)(x_i + y_i) + v_i - I_i]$ being maximized is that the n households with the largest $(b + c)(x_i + y_i) - I_i$ adopt PV systems. This is accomplished with $p = b + c$, $r = b + c$, and $s = g(b + c, b + c)$ because only n adopters satisfy $(b + c)(x_i + y_i) + v_i + g(b + c, b + c) \geq I_i$. If FITs are applied to all PV electricity, this combination is always feasible by considering $r = p$ and $g(b+c, b+c) = f(b+c)$. In fact, social welfare was maximized with $p = b + c$, $s = f(b + c)$ in Sect. 5.4.3. In contrast, if FITs are applied to surplus PV electricity, that combination is not feasible due to constraint (5.7), unless b is so small that $b \approx 0$ holds. Therefore, the maximized social welfare in the case of FITs applied to all PV electricity is larger than that in the case of FITs applied to surplus PV electricity.

5.6 The Model for the Business Sector

In this section, the model for the business sector is developed to work out an optimal combination of FITs and capital subsidies. The variables in the model are essentially the same as those presented so far in the model for the residential sector except for the installed capacity of PV panels. The model is similar to the model of FITs for the business sector in Chap. 3. We first set up the model with a brief explanation of variables and then examine the social welfare maximization problem.

5.6.1 Decision-Making of a Firm

A firm has many potential sites for PV generation, each of which is indexed by i. Assume that site i has a potential installed PV capacity of m_i, which is expressed in kW. A unit of PV capacity at site i can generate the amount of PV electricity of z_i during its lifetime. A discount rate is taken into account so that the firm can obtain the payment pm_iz_i over the lifetime of the site if PV electricity is sold at a price p. If the firm installs a PV system at site i, it incurs the cost of I_i per unit of PV capacity. The discount rate is also taken into account for I_i. Thus, the total cost to invest in PV generation at site i is m_iI_i.

To attain M kW of PV capacity in the current term, a government offers a price p per kWh at which a firm can sell PV electricity, as well as a capital subsidy s per kW of PV capacity, which a firm can be granted up front for each site. Accordingly, the PV system at site i yields revenue of $pm_iz_i + sm_i$ and incurs the cost of m_iI_i.

Therefore, the firm develops site i if and only if

$$pz_i + s - I_i \geq 0. \tag{5.9}$$

Then, the straight line $s = -pz_i + I_i$ forms a boundary in the ps–plane between site i being developed and not being developed.

5.6.2 Potential Combinations of FITs and Capital Subsidies

The government aims at having M kW of PV capacity installed. However, it is difficult to make the total installed capacity exactly equal to M kW through the combination of FITs and capital subsidies. Generally, the PV capacity to be installed at a site is determined by exogenous conditions that are beyond a developer's decision-making, such as availability of land area, restrictions on connection to the power grid, and so forth. Thus, it is assumed that if a site is developed, all of the PV electricity generated at the site is eligible for FITs and capital subsidies.

Accordingly, if the government offers a combination of FITs and capital subsidies, the marginal site—the first site that would no longer satisfy condition (5.9) if FIT or capital subsidy levels were infinitesimally decreased—will generally cause the total installed capacity to be larger than M kW. Hence, it is assumed that the government seeks to attain at least M kW of PV capacity in that if the marginal site were no longer developed, the total installed capacity would be strictly smaller than M kW. Taking this into account, we will simply state that the government aims to attain M kW of installed PV capacity.

The pair (p, s) that causes M kW of PV capacity to be installed may be identified as follows. First, for every site i, draw a straight line $s = -pz_i + I_i$ in the ps–plane. Second, find the site, designated as i_j, that has the jth lowest I_i on the s–axis and identify the k such that $\sum_{j=1}^{k-1} m_{i_j} < M$ and $\sum_{j=1}^{k} m_{i_j} \geq M$. Site i_k is the marginal site for $p = 0$ and sufficiently small ps. Define d_{i_k} as $d_{i_k} = M - \sum_{j=1}^{k-1} m_{i_j}$. In other words, d_{i_k} of m_{i_k} kW contributes essentially to reaching the target capacity of M kW. In contrast, $\left(m_{i_k} - d_{i_k}\right)$ kW are installed beyond M kW. Third, starting at $\left(0, I_{i_k}\right)$ on the s–axis, trace the line $s = -pz_{i_k} + I_{i_k}$ toward lower right until it intersects with another line.

It is somewhat complicated to determine which line should be chosen when an intersection is encountered. Consider the intersection of two lines, $s = -pz_i + I_i$ and $s = -pz_j + I_j$, where site i is the marginal site before the intersection. There are two cases of intersection between the two lines. The first case is where $z_i > z_j$ (Fig. 5.1a). In this case, site j is also developed before the intersection. If $m_j < m_i - d_i$, site i is still the marginal site, and site j is no longer developed after the intersection. Thus, trace the same line $s = -pz_i + I_i$ after the intersection as before. Note that the total installed capacity decreases by m_j in this case. If $m_j \geq m_i - d_i$, site j becomes the marginal site, while site i is still developed but is no longer the marginal site. Thus, trace the other line $s = -pz_j + I_j$ after the intersection. Note that in this case, the total installed capacity does not change.

The second case is where $z_i < z_j$ (Fig. 5.2b). Site j is not developed before the intersection because site i is the marginal site. Suppose $m_j \geq d_i$. After the intersection, site j becomes the marginal site, while site i is no longer developed. Thus, trace $s = -pz_j + I_j$ after the intersection. Note that the total installed capacity decreases by $m_i - m_j$ if $m_j < m_i$ and increases by $m_j - m_i$ if $m_j \geq m_i$. Next, suppose $m_j < d_i$. After the intersection, site i is still the marginal site, with site j being developed as well. Thus, trace the same line $s = -pz_i + I_i$ after the intersection as before. Note that the total installed capacity increases by m_j in this case.

Repeating this procedure, we can form the broken line $s = f(p)$ for the business sector. In other words, the government offers PV electricity price p and capital subsidy $s = f(p)$ to the business sector to ensure that M kW of PV capacity is installed in the current term. In the next section, we investigate the optimal p.

5.6.3 Social Welfare Maximization for the Business Sector

In this subsection, we address the social welfare maximization problem. Before doing this, we briefly discuss the maximization problem of PV electricity and the minimization problem of promotion cost. First, the maximization problem of PV electricity is not clearly solved because whether a site is developed, which is determined by inequality (5.9), is independent of the PV capacity, m_i, of that site. Second, the minimization problem of promotion cost is also not clearly solved: recall that this problem was not clearly solved for the residential sector in Sect. 5.4.2.

Let Ω_p again be the set of sites that are developed when the government offers p and $s = f(p)$ to ensure that M kW of PV capacity is installed. The problem of social welfare maximization is

$$\text{maximize} \sum_{i \in \Omega_p} m_i \left[(b + c)z_i - I_i\right]. \tag{5.10}$$

The solution to problem (5.10) is again $p = b+c$, which yields $s = f(b+c)$: from inequality (5.9), $(b + c)z_i - I_i \geq -s$ for every site $i \in \Omega_{b+c}$ and $(b + c)z_i - I_i < -s$ for any site $i \notin \Omega_{b+c}$. Any p close to $b + c$ yields the same maximum as long as $\Omega_p = \Omega_{b+c}$.

5.7 Discussion

Let us consider the implications of the results for the residential sector in terms of the previous studies mentioned in Sect. 5.2. First, our results are consistent with the results of Mir-Artigues and del Río (2014), namely that financial costs are invariable—for any combination of FITs and capital subsidies—if an investor is to achieve a given level of profitability. In our context, the combination of FITs and capital subsidies that causes an individual household to adopt a PV system is represented by a straight line (5.1). This means that any combination on that line provides a household with a fixed amount of welfare.

Second, our investigation does not provide a clear result, while Hsu (2011) showed that the policy cost will be smaller with lower FITs and higher capital subsidies. According to our model analysis, it is not clear whether the promotion cost increases or decreases. This will be further investigated by means of a simulation in the next Chap. 6.

Third, to maximize social welfare when FITs are applied to all PV electricity, FITs should be set at the avoided cost per unit of PV electricity, whereas capital subsidies should be set to control the number of adopters. This also holds for the business sector. The result is consistent with that of Oi (1971), implying that FITs are used for efficiency and capital subsidies are for participation. Interestingly, FITs work for efficiency, although a household or a firm can only decide whether to adopt a PV system or develop a site and cannot control the amount of PV electricity to be generated. The result also supports the discussion, deduced from Klein (2008), that FITs may correspond to the avoided-cost approach and that capital subsidies may correspond to the RES-E generation-cost approach.

In contrast, if FITs are applied to surplus PV electricity in the residential sector, the above principle, namely that FITs should be set at the avoided cost per unit of PV electricity, is—to some extent—distorted theoretically.

5.8 Conclusion

This chapter shed light on the combined use of FITs and capital subsidies to encourage households to adopt PV systems or firms to develop a site for PV generation. Optimal combinations were examined using a microeconomic model that drew on the idea of two-part tariffs. This study not only provides practical guidelines on the setting of FIT and capital subsidy levels but also enhances understanding of the FIT and capital subsidy mechanisms through a theoretical investigation.

The findings, which are directly connected to policy implications, are summarized as follows. First, suppose that FITs are applied to all PV electricity. The amount of PV electricity to be generated is larger with higher FITs and lower capital subsidies if we assume the installed capacity is the same among adopters in the residential sector; the combination that minimizes the promotion cost is unclear but depends on the parameter values. Social welfare is maximized if FITs are set at the avoided cost per unit of PV electricity, with capital subsidies being used to control the number of adopters in the residential sector or the PV capacity developed in the business sector. Note that the avoided cost includes not only the expenses of conventional power generation but also the external costs.

Second, suppose that FITs are applied to surplus PV electricity for the residential sector. The FIT level that maximizes social welfare is not clear. What is clear is that the maximized social welfare is smaller than it would be if FITs were applied to all PV electricity.

These results for the residential sector will be further investigated with a simulation in the next Chap. 6.

The scope of the present study is limited to the topic of the adoption of PV systems, but the approach employed may be applied to the adoption of wind-power generation systems. Then, the main results would still hold because the underlying structure is universal with regard to combined FITs and capital subsidies as an application of two-part tariffs.

This study investigated social welfare maximization. This enabled us, on the one hand, to examine the efficiency of a combined FIT and capital subsidy system. On the other hand, the problem of equity, that is, how to finance the cost of these support measures, was not explicitly addressed. There are two questions to answer in this regard. First, who pays the cost? It should be noted that generating PV electricity might be considered to provide public goods such as alleviation of global warming and improved energy security. Hence, it could be argued that every household should contribute to supplying PV electricity to society. The second question is how the cost should be allocated among all households. It is often asserted that these support measures are a type of wealth transfer from those who cannot afford to adopt PV systems to those who can afford PV adoption. This problem may be resolved by carefully designing a payment vehicle. If a support measure is financed by tax revenues, a progressive taxation system such as an income tax may help resolve the problem to some extent. Alternatively, financing through increased electricity rates could be somewhat rationalized because richer households are more likely to consume higher amounts of electricity. In our model, FITs are financed through increased electricity rates, while capital subsidies are financed by taxes. The problem of equity is so sensitive that deeper explorations should be required, separate from the present study.

Although this study focuses on the earlier stage of RET diffusion, it also suggests a possible method for pricing RES-E once a diffusion goal is achieved. According to the results of social welfare maximization, as the diffusion of RETs progresses, the role of capital subsidies becomes smaller and may sometimes be negative. In this event, the negative value of capital subsidies would play the role of an access charge to the power grid: a customer-generator or distributed generator must pay a lump sum charge for the right to sell RES-E to the grid at a FIT. Furthermore, it is necessary to reconsider the avoided costs on which FITs are based. As more distributed generators are connected to the grid, the power system will become so complicated that preserving system security, for example, voltage stability, will be more difficult. This may be regarded as a cost of the system and should be considered when a regulatory authority sets FIT levels. Given the necessity and likelihood of further RET diffusion, additional research into pricing mechanisms is therefore essential.

Appendix 1: Proof of the Lemma for the Social Welfare Maximization Problem

First, let us verify that for a given p, $\sum_{i \in \Theta_{p,c,g(p,c)}} [c(x_i + y_i) - I_i]$ is maximized at $p = c$. The same reasoning may be applied here as in Sect. 5.4.3 (see also Fig. 5.2b). Suppose that the marginal household changes from i to j when p increases to p_1, subject to $r = c$. Then, $\sum_{i \in \Theta_{p,c,g(p,c)}} [c(x_i + y_i) - I_i]$ changes if and only if the marginal household's surplus PV electricity increases, i.e., $x_i < y_j$. Otherwise, the set $\Theta_{p,c,g(p,c)}$ of adopters remains the same, and the sum does not change. Plugging $p = c$ into $s = -px_i - cy_i + I_i$ from Eq. (5.5) yields $-cx_i - cy_i + I_i > -cx_j - cy_j + I_j$, or $c(x_i + y_i) - I_i < c(x_i + y_i) - I_j$ if $p_1 < c$ and that $-cx_i - cy_i + I_i < -cx_j - cy_j + I_j$, or $c(x_i + y_i) - I_i > c(x_j + y_j) - I_j$ if $p_1 > c$. Because the adopters, other than the marginal households i or j, are the same, $\sum_{i \in \Theta_{p,c,g(p,c)}} [c(x_i + y_i) - I_i]$ is maximized at $p = c$.

Next, let us verify the second half of the lemma. Plugging $p = 0$ into $s = -px_i - cy_i + I_i$ and $s = -px_j - cy_j + I_j$ reveals that $\sum_{i \in \Theta_{p,c,g(p,c)}} (cy_i - I_i)$ decreases if $x_i < x_j$. Hence, while $\sum_{i \in \Theta_{p,c,g(p,c)}} x_i$ increases, it is not possible to state how $\sum_{i \in \Theta_{p,c,g(p,c)}} y_i$ changes; this depends on the parameter values. Thus, changes in $\sum_{i \in \Theta_{p,c,g(p,c)}} b(x_i + y_i)$ also depend on the parameter values.

References

Cerdá E, del Río P (2015) Different interpretations of the cost-effectiveness of renewable electricity support: some analytical results. Energy 90:286–298

del Río P, Cerdá E (2014) The policy implications of the different interpretations of the cost-effectiveness of renewable electricity support. Energy Policy 64:364–372

del Río P, Mir-Artigues P (2014) Combinations of support instruments for renewable electricity in Europe: A review. Ren Sustain Energy Rev 40:287–295

Dusonchet L, Telaretti E (2010) Economic analysis of different supporting policies for the production of electrical energy by solar photovoltaics in western European Union countries. Energy Policy 38:3297–3308

Hirvonen J, Kayo G, Cao S, Hasan A, Sirén K (2015) Renewable energy production support schemes for residential-scale solar photovoltaic systems in Nordic conditions. Energy Policy 79:72–86

Hsu C-W (2011) Using a system dynamics model to assess the effects of capital subsidies and feed-in tariffs on solar PV installations. Appli Energy 100:205–217

Klein A (2008) Feed-in tariff designs: options to support electricity generation from renewable energy sources. VDM Verlag Dr, Müller, Saarbrücken

Lesser JA, Su X (2008) Design of an economically efficient feed-in tariff structure for renewable energy development. Energy Policy 36:981–990

Mayr D, Schmidt J, Schmid E (2014) The potentials of a reverse auction in allocating subsidies for cost-effective roof-top photovoltaic system deployment. Energy Policy 69:555–565

METI (2013) The report on the diffusion of PV systems in Japan. www.meti.go.jp/meti_lib/report/2013fy/E002502.pdf. Accessed 6 Jul 2017 (In Japanese)

Mir-Artigues P, del Río P (2014) Combining tariffs, investment subsidies and soft loans in a renewable electricity deployment policy. Energy Policy 69:430–442

Mir-Artigues P, del Río P (2016) The economics and policy of solar photovoltaic generation. Springer, Cham

Oi WY (1971) A Disneyland dilemma: two-part tariffs for a Mickey Mouse monopoly. Q J Econ 85:77–96

Oliva S, MacGill I, Passey R (2014) Estimating the net societal value of distributed household PV systems. Sol Energy 100:9–22

Pirnia M, Nathwani J, Fuller D (2011) Ontario feed-in tariffs: system planning implications and impacts on social welfare. Electr J 24:18–28

Yamamoto Y (2012) Pricing electricity from residential photovoltaic systems: a comparison of feed-in tariffs, net metering, and net purchase and sale. Sol Energy 86:2678–2685

Zhao Y, Tang KK, Wang L (2013) Do renewable electricity policies promote renewable electricity generation? Evidence from panel data. Energy Policy 62:887–897

Chapter 6
Simulations of a Combination of Feed-in Tariffs and Capital Subsidies

Abstract In Chap. 5, we studied optimal combinations of feed-in tariffs (FITs) and capital subsidies. The theoretical investigation we developed through our microeconomic model provided some rigorous results as well as some unclear results. The purpose of this chapter is to confirm the results and develop new insights into those that were unclear in Chap. 5. To this end, we conduct simulations of the adoption of photovoltaic (PV) systems in the residential sector. At the same time, these simulations will help to clarify the theoretical model. Parameter values are set based on various sources of data. The simulations presented here have verified the theoretical results yielded in Chap. 5 and provided new findings with regard to the amount of PV electricity, the promotion cost, and social welfare. In particular, the promotion cost, which had been unclear in Chap. 5, showed a U–shaped curve, suggesting that there is a FIT level that minimizes the promotion cost. The comparison between the two cases, FITs for all PV electricity and FITs for surplus PV electricity, provided some reasonable results. Our simulation is intended to supplement the theoretical investigation in Chap. 5; it is not intended as an empirical investigation. More in-depth simulations should be conducted to obtain practical guidelines on setting FIT and capital subsidy levels.

Keywords Simulation · Feed-in tariff · Capital subsidy
Residential photovoltaic system

6.1 Introduction

In Chap. 5, we developed a model to investigate a combination of feed-in tariffs (FITs) and capital subsidies, and those model investigations revealed some important features of the combinations. However, some features could not be clarified. The

---

Modified, with permission of Elsevier, from Yamamoto, Y., Feed-in tariffs combined with capital subsidies for promoting the adoption of residential photovoltaic systems, Energy Policy, 111, 312–320, Elsevier, 2017. I would like to thank Elsevier.

Y. Yamamoto, *Feed-in Tariffs and the Economics of Renewable Energy*,
https://doi.org/10.1007/978-3-319-76864-9_6

purpose of this chapter is to confirm the results obtained in Chap. 5 and to obtain some insights into results that could not be clarified. We focus particularly on the results regarding the adoption of a PV system in the residential sector, which was predominant investigation of Chap. 5. It should be noted that the simulation in this chapter is aimed at improving understanding of the results in Chap. 5; it is not intended as an empirical investigation, which is beyond the scope of this chapter.

The main results, which are obtained for the residential sector, are once again summarized here. Suppose FITs are applied to all PV electricity. The amount of PV electricity to be generated is larger with higher FITs and lower capital subsidies; the combination that minimizes the promotion cost is unclear and depends on the parameter values. Social welfare is maximized when FITs are set at the avoided cost per unit of PV electricity, with capital subsidies being used to control the number of adopters. If it is supposed that FITs are applied to surplus PV electricity, the FIT level that maximizes social welfare is not clear; what is clear is that the maximized social welfare is smaller than it would be if FITs were applied to all PV electricity.

The remainder of the chapter is organized as follows. In Sect. 6.2, the model in Chap. 5 is briefly reviewed to clarify what type of model is used for simulation. Parameter values are set in Sect. 6.3. In Sect. 6.4, simulation results are presented, first in the case of FITs applied to all PV electricity and then in the case of FITs applied to surplus PV electricity. Finally, Sect. 6.5 concludes.

6.2 The Model Used for Simulations

We use the model developed in Chap. 5 for this chapter's simulations. To begin, let us briefly review the model.

6.2.1 Definitions of Variables

A government that aims to have n of N households adopt photovoltaic (PV) systems implements FITs and capital subsidies in combination. On the one hand, the government sets a FIT level at p; an adopter can sell PV electricity generated at price p during a set period of years, T, which is, for the sake of simplicity, equal to the lifetime of a PV system. The FITs are applied to either all PV electricity or surplus PV electricity, where surplus PV electricity is defined as PV electricity that is not self-consumed and is fed into the power grid. On the other hand, the government grants a capital subsidy, s, to an adopter; s is constant because the rooftop area that can be used for installing a PV system may be similar among houses. The government finances the cost of FITs by raising the retail electricity rate as a regulator of an electric utility, while the cost of capital subsidies is collected through taxation.

For a household i, I_i is the cost of a PV system, z_i is the amount of PV electricity to be generated by a PV system during the period for FITs, and v_i is the satisfaction adopter i experiences from the PV system.

A discount rate, ε, is taken into account for z_i so that an adopter i can obtain the payment pz_i if it can sell all of the PV electricity generated throughout the period. In other words, z_i in the model is the value already discounted by ε.

Household i's amount of PV electricity, z_i, is divided into the amount not self-consumed, namely the surplus, x_i, and the amount self-consumed, y_i, where $x_i - y_i = z_i$. The PV electricity not self-consumed is fed into the grid to sell at p.

Lastly, b is the benefit of the avoided external costs realized per unit of PV electricity, and c is a unit cost of conventional electricity generation. The retail electricity rate is r, which is equal to c before FITs are implemented.

6.2.2 The Structure of the Model

The investigation is divided into two cases in terms of which part of PV electricity is eligible for FITs. First, suppose that all of the PV electricity is eligible for FITs. Given a combination of p and s, a household i adopts a PV system if and only if $pz_i + v_i + s \geq I_i$. A straight line

$$s = -pz_i - v_i + I_i \quad (p \geq 0) \tag{6.1}$$

forms a boundary in the ps–plane between i's adopting and not adopting.

Plotting the nth highest s against p forms a curve, $s = f(p)$, in the ps–plane, which is the locus of the potential combinations of p and s. Let Ω_p be the set of n adopters. Then, the amount of PV electricity, the promotion cost, and the social welfare are, respectively,

$$\sum_{i \in \Omega_p} z_i, \tag{6.2}$$

$$(p - c) \sum_{i \in \Omega_p} z_i + nf(p), \tag{6.3}$$

$$\sum_{i \in \Omega_p} [(b + c)z_i + v_i - I_i]. \tag{6.4}$$

Second, suppose that only surplus PV electricity is eligible for FITs. Given a combination of p, r, and s, a household i adopts a PV system if and only if $px_i + ry_i + v_i + s \geq I_i$. A plane

$$px_i + ry_i + v_i + s - I_i = 0 \quad (p \geq 0, r \geq 0) \tag{6.5}$$

forms a boundary in the prs–space between i's adopting and not adopting.

Two conditions must be satisfied. First, exactly n households adopt PV systems. Second, the cost of FITs is raised through increased electricity rates. From these conditions, r and s may be expressed as functions of p, i.e., $r(p)$ and $s(p)$, respectively. Let Θ_p be the set of n adopters. Then, the amount of PV electricity, the promotion cost, and the social welfare are, respectively,

$$\sum_{i \in \Theta_p} (x_i + y_i), \tag{6.6}$$

$$(p - c) \sum_{i \in \Theta_p} x_i + (r(p) - c) \sum_{i \in \Theta_p} y_i + ns(p), \tag{6.7}$$

$$\sum_{i \in \Theta_p} [(b + c)(x_i + y_i) + v_i - I_i]. \tag{6.8}$$

It should be noted that because self-consumption of PV electricity is considered to be sold and purchased at the retail electricity rate by an adopter, it is included in the promotion cost calculation (6.7), compared with (6.3). In other words, the electric utility procures the amount of self-consumed electricity at $r(p)$ instead of generating it on its own at a unit cost of c.

6.3 Setting of Parameter Values

Parameter values are set in the same way as in Chap. 4 (Table 4.2). They are again summarized in Table 6.1: the conventional power generation cost $c = \$0.11/\text{kWh}$ (EIA 2017); the social benefit per unit of PV electricity $b = \$0.1/\text{kWh}$ in terms of CO_2 emissions reduction (0.5 $\text{kgCO}_2/\text{kWh} \times \$20/\text{tCO}_2$: JPEA 2015; IEA 2012); and the fixed sum of electricity consumption $L = 1060$ GWh during the lifetime of a PV system $T = 20$ years (5300 kWh/year × 10,000 households × 20 years: JPEA 2015).

A PV system cost, I_i, is drawn from a normal distribution, with the mean of \$14,000 (METI 2013) and standard deviation of \$1000.

A valuation of a household on a PV system, v_i, is drawn from a normal distribution with a mean of \$500 and a standard deviation of \$200.

Household i's amount of PV electricity, z_i, to be generated during the lifetime of a PV system, is drawn from a normal distribution, with a mean of 80,000 kWh ($4400\ \text{kWh/year} \times (1 - 0.99^{20})/(1 - 0.99)$: NEDO 2017) and standard deviation of 16,000 kWh. The discount rate is set at $\varepsilon = 0.01$. Household i's self-consumption of electricity, y_i, is constructed from the z_i multiplied by a random variable drawn from a uniform distribution ranging from 0.2 to 0.4. Finally, the surplus electricity, x_i, is computed as $x_i = z_i - y_i$.

The target number of adopters $n = 500$ in relation to the total number of households (detached houses), $N = 10{,}000$.

Table 6.1 Parameter values

I_i	Household i's cost of a PV system	Drawn from a normal distribution (mean: \$14,000, SD: \$1000)
v_i	Household i's satisfaction obtained from adopting a PV system	Drawn from a normal distribution (mean: \$500, SD: \$200)
z_i	Household i's amount of PV generation	Drawn from a normal distribution (mean: 80,000 kWh, SD: 16,000 kWh)
x_i	Household i's amount of surplus PV electricity	$z_i - y_i$
y_i	Household i's amount of self-consumed PV electricity	z_i multiplied by a random variable drawn from a uniform distribution ranging from 0.2 to 0.4
b	Benefit of the avoided external cost per unit of PV electricity	\$0.01/kWh
c	Unit cost of conventional power generation	\$0.11/kWh
N	Total number of households	10,000
n	Target number of adopters	100
L	Electricity consumption by all households	1060 GWh
T	Lifetime of a PV system	20 years
ε	Discount rate	0.01

6.4 Results and Discussion

Simulation results are presented in this section. The results presented in Sect. 6.4.1 suppose that FITs are applied to all PV electricity. On the other hand, the results presented in Sect. 6.4.2 suppose that FITs are applied to only surplus PV electricity.

6.4.1 The Results with FITs Applied to All PV Electricity

Suppose that FITs are applied to all PV electricity. The subsidy $s = f(p)$ forms an almost straight line, with a downward slope in the *ps*–plane (Fig. 6.1). It can be observed that $f(p)$ is negative if $p > 0.1232$.

The total amount of PV electricity increases with decreasing rates of change (Fig. 6.2). For a lower p, an increment of p significantly induces the households that will potentially generate a large amount of PV electricity to adopt PV systems. As p becomes higher, this effect becomes weak, although the total amount of PV

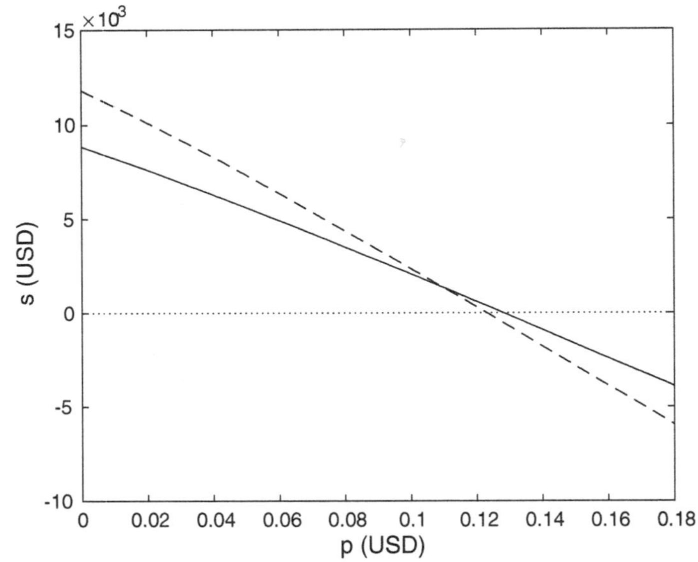

Fig. 6.1 The subsidy. The dashed curve represents the subsidy in the case of FITs for all PV electricity, $f(p)$; the solid curve represents the subsidy in the case of FITs for surplus PV electricity, $s(p)$. The shapes of the two curves are almost straight, with downward sloping. However, the slope of $s = s(p)$ is slightly gentler than that of $s = f(p)$

electricity continues to increase. This simulation verifies the result yielded by the theoretical investigation in Chap. 5.

The promotion cost decreases suddenly at first, but then it increases (Fig. 6.3). This U–shaped curve may be attributable to the increase in the amount of PV electricity (Fig. 6.2); it is suggested that as the amount of PV electricity increases, the cost of purchasing PV electricity has more effects on the promotion cost.

It is confirmed that social welfare is maximized when $p = b + c = 0.12$, with a maximum of $323,390. In addition, the same maximum can be attained by the p ranging from 0.1193 to 0.1207 (Fig. 6.4). This is consistent with the theoretical results in Chap. 5.

6.4.2 The Results with FITs Applied to Surplus PV Electricity

Next, suppose that FITs are applied to surplus PV electricity. The subsidy $s(p)$ is shown in the ps–plane in Fig. 6.1. It is roughly a straight line but gentler than $s = f(p)$; on the one hand, for a lower p, the surcharged electricity rate r is higher than p so that $px_i + ry_i > p(x_i + y_i) = pz_i$. Hence, $s(p) < f(p)$ holds. On the other hand, for a higher p, r is lower than p so that $px_i + ry_i < pz_i$. Accordingly, $s(p) > f(p)$ holds. Therefore, $s(p)$ is gentler than $f(p)$.

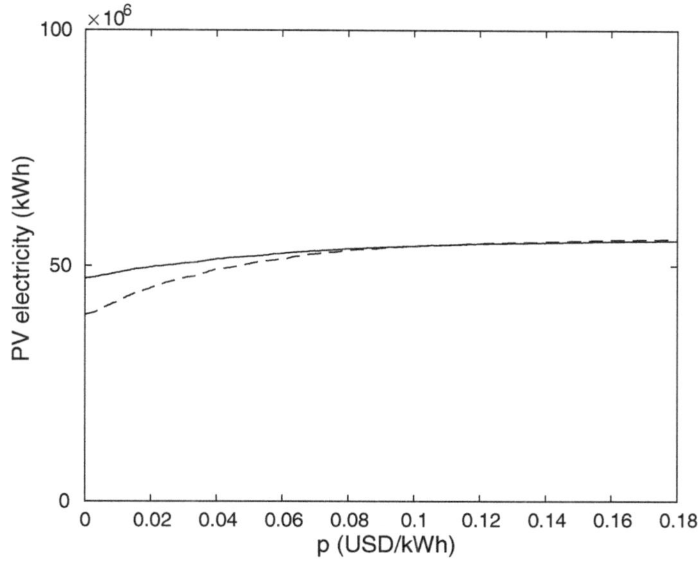

Fig. 6.2 The amount of PV electricity. The dashed curve corresponds to FITs for all PV electricity, and the solid curve corresponds to FITs for surplus PV electricity. The amount of PV electricity increases with decreasing rates of change in both cases. When p is low, the amount of PV electricity is slightly larger with FITs for surplus PV electricity than with FITs for all PV electricity

The surcharged electricity rate $r(p)$ is shown in Fig. 6.5. It is confirmed that $r(p) \approx c = 0.11$; the shape of $r(p)$ is almost horizontal with the slope being approximately 0.03. In other words, the amount of the surcharge is relatively small. This may be attributable to the assumption that the investigation concerns the earlier stage of the diffusion of PV systems: the target number of adopters, $n = 500$, is much smaller than the total number of households, $N = 10{,}000$.

The shape of the amount of PV electricity is similar to that of FITs for all PV electricity (Fig. 6.2); it increases with decreasing rates of change. Taking a closer look at Fig. 6.2, we notice that for a lower p, PV electricity is larger than in the case of FITs for all PV electricity. The reason is that for such ps, it is more likely that households expecting to generate a larger amount of PV electricity are willing to adopt in the case of FITs for surplus PV electricity because $r(p)$ is higher than p.

The shape of the promotion cost is also similar to that of FITs for all PV electricity (Fig. 6.3). It decreases at first, but then increases. Taking a closer look at Fig. 6.3, we notice that for a lower p, the promotion cost is smaller than in the case of FITs for all PV electricity. This may be explained as follows. If FITs are applied to all PV electricity, a larger capital subsidy will be required to attain the target number of adopters when p is low. In contrast, if FITs are applied only to surplus PV electricity, the target number of adopters may be attained with a smaller capital subsidy because part of PV electricity is priced at $r(p)$, which is higher than p. The effect of a larger

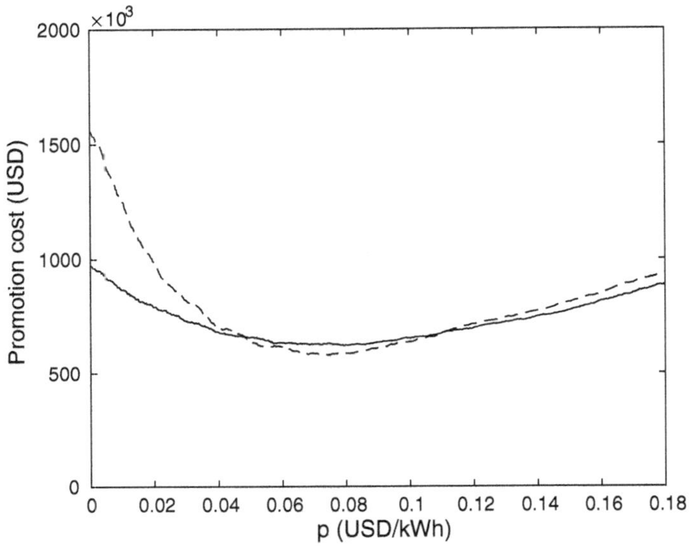

Fig. 6.3 The promotion cost. The dashed curve corresponds to FITs for all PV electricity, and the solid curve corresponds to FITs for surplus PV electricity. The promotion cost decreases at first but then increases in both cases. When p is low, the promotion cost is apparently larger with FITs for all PV electricity than with FITs for surplus PV electricity

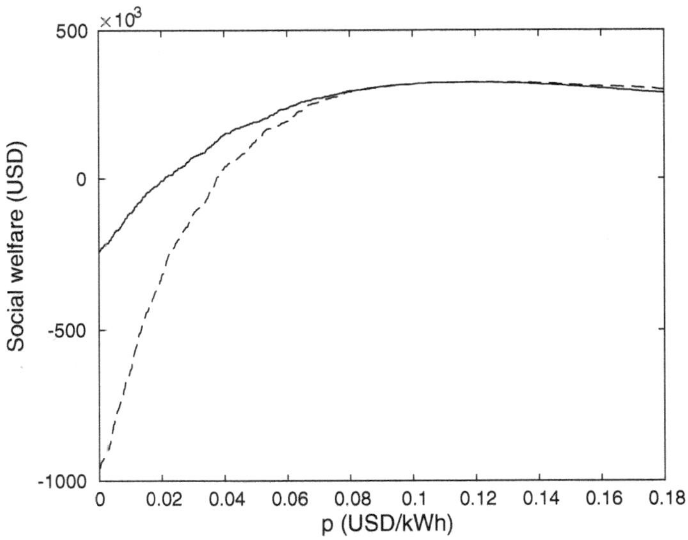

Fig. 6.4 Social welfare. The dashed curve corresponds to FITs for all PV electricity, and the solid curve corresponds to FITs for surplus PV electricity. Social welfare is maximized at $p = 0.12$ and in its neighborhood in the case of FITs for all PV electricity. On the other hand, in the case of FITs for surplus PV electricity, the maximizer is smaller than that in the case of FITs for all PV electricity

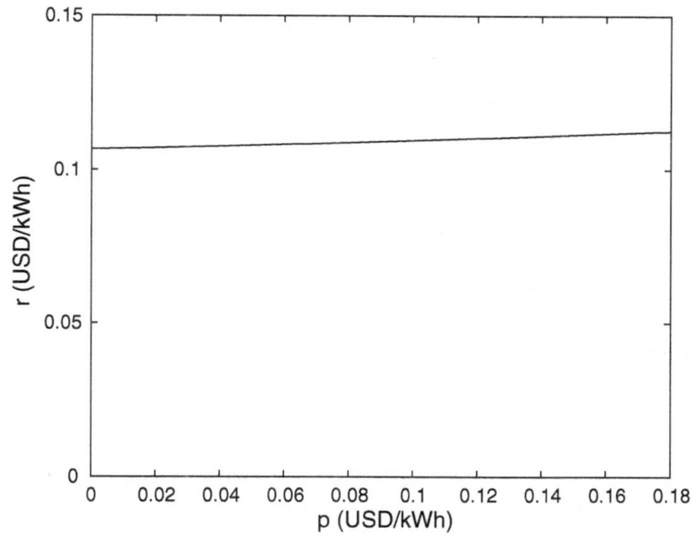

Fig. 6.5 The surcharged electricity rate in the case of FITs for surplus PV electricity. The surcharged electricity rate is almost constant at $r = 0.11$

capital subsidy on the promotion cost would be stronger than the effect of a partly higher purchase price of PV electricity.

This result with regard to the promotion cost agrees in part with Hsu's (2011) results that the policy cost will be smaller with lower FITs and higher capital subsidies. Our simulation shows that although it decreases for lower FITs, the promotion cost predominantly increases as the FIT level becomes high.

The maximized social welfare is \$322,550 at the p ranging from 0.1180 to 0.1188 (Fig. 6.4). This figure is slightly smaller than the maximum in the case of FITs for all PV electricity, which is \$323,390. This confirms the result of Sect. 5.5.2 in Chap. 5 that the maximized social welfare in the case of FITs for surplus PV electricity is smaller than that in the case of FITs for all PV electricity. However, it should be noted that the difference is so small that whether FITs are applied to either all or surplus PV electricity will not matter much, given the current set of parameter values. Furthermore, the p that maximizes social welfare, called the maximizer, is less than $b + c = 0.12$ and is close to the standard electricity retail rate $c = 11$. To be specific, the maximizer p, ranging from 0.1180 to 0.1188, is smaller than the maximizer p, ranging from 0.1193 to 0.1207, in the case of FITs for all PV electricity in Sect. 6.4.1. These results regarding social welfare maximization are consistent with the theoretical results in Chap. 5.

6.5 Conclusion

In this chapter, simulations were conducted to confirm the results we obtained from the theoretical investigation in Chap. 5.

The simulation results are consistent with those obtained in Chap. 5. The amount of PV electricity increases as the FIT level becomes high. Although the theoretical investigation in Chap. 5 did not provide a clear result about how the promotion cost varies, the simulation suggests that it forms a U–shaped curve. The results regarding social welfare maximization are consistent with the theoretical investigation in Chap. 5. The comparison between the two cases, FITs applied to all PV electricity and FITs applied to surplus PV electricity, reveals rational results, which are also deduced from the theoretical investigation.

It should be noted that the promotion cost has a U–shaped curve, suggesting that this cost may be minimized at a certain level of FITs. It is not clear whether this result is robust, that is, whether it still holds if another set of parameter values is used. We will address this problem in future work.

The simulation presented in this chapter offers a suggestion regarding social welfare maximization. When FITs are applied to surplus PV electricity, the FIT level that maximizes social welfare becomes close to the electricity retail rate prior to the addition of a surcharge. This may imply that net metering may be an option for maximizing social welfare: net metering allows a customer-generator to both sell and purchase electricity at the standard electricity retail rate. If net metering is applicable, a government need not determine a FIT level administratively. Considering the relative ease of net metering, the government may opt to use this approach to social welfare maximization.

Recall that the numerical example in this chapter is aimed at helping to clarify the theoretical results in Chap. 5; the intention is not to conduct an empirical investigation. In-depth simulations, intended to provide practical guidelines on setting FIT and capital subsidy levels, should be conducted to ensure the effective diffusion of renewable energy technologies.

References

EIA (2017) Sources & users. www.eia.gov/electricity/sales_revenue_price/pdf/table6.pdf. Accessed 26 Feb 2017

Hsu C-W (2011) Using a system dynamics model to assess the effects of capital subsidies and feed-in tariffs on solar PV installations. Appl Energ 100:205–217

IEA (2012) World energy outlook 2012. IEA, Paris

JPEA (2015) A leaflet of residential PV generation. www.jpea.gr.jp/pdf/susume_pamphlet.pdf. Accessed 26 Feb 2017 (in Japanese)

METI (2013) The report on the diffusion of PV systems in Japan. www.meti.go.jp/meti_lib/report/2013fy/E002502.pdf. Accessed 6 Jul 2017 (in Japanese)

NEDO (2017) MONSOLA-11. app0.infoc.nedo.go.jp/metpv/monsola.html. Accessed 7 Jul 2017

Chapter 7
The Model with Continuous Variables

Abstract In Chap. 5, we developed a microeconomic model to investigate optimal combinations of feed-in tariffs (FITs) and capital subsidies for the adoption of solar photovoltaic (PV) power generation systems in the residential sector. In that model, it was assumed that a household, a potential adopter, is characterized by several variables related to PV generation. Then, we determined the levels of FIT and capital subsidy at which a household would adopt a PV system. Accordingly, we can determine combinations of FITs and capital subsidies at which a certain number of households adopt. However, in reality, whether a particular household adopts may make little sense; what a government is concerned with is the total number of adopters, and it will control FIT and capital subsidy levels so that the target amount of adoption may be achieved. Hence, in this chapter, we develop a model constructed with continuous variables: the variables that characterize a household take continuous values so that a government considers solely the amount of adoption. By using this model, we consider three optimality criteria: maximization of PV electricity, minimization of promotion cost, and maximization of social welfare. The same results are obtained as in Chap. 5 with respect to the optimal combinations of FITs and capital subsidies.

Keywords Feed-in tariff · Capital subsidy · Continuous variable · Social welfare

7.1 Introduction

In Chap. 5, we developed a microeconomic model for the residential and business sectors to examine optimal combinations of feed-in tariffs (FITs) and capital subsidies for promoting the adoption of photovoltaic (PV) systems. In the model for the residential sector, each household is characterized by several variables: the amount of PV electricity to be generated, the cost of a PV system, and the valuation on PV generation; if a government offers a combination of FITs and capital subsidies, some households adopt PV systems and others do not. The government can thereby control the number of adopters by setting the FIT and capital subsidy levels. In this

Y. Yamamoto, *Feed-in Tariffs and the Economics of Renewable Energy*,
https://doi.org/10.1007/978-3-319-76864-9_7

sense, such a model can be considered a model with discrete variables. This way of modeling is applied to the business sector as well.

Considering that the target number of adopters a government sets is significantly large, as is the total number of households, we propose an alternative way of modeling wherein a household is characterized by continuous variables. This way of modeling may be so common in economic literature that quantity x of a good is usually expressed by a continuous variable as $x \geq 0$ rather than a discrete variable as $x = 0, 1, 2, \ldots$. The purpose of this chapter is to present a continuous-variable version of the model for investigating a combination of FITs and capital subsidies. Modeling with continuous variables will facilitate understanding of the model investigation, although mathematical rigor might be lost to some extent.

The modeling presented in this chapter may also provide another way of investigating decision-making. Recall that in the model in Chap. 5, we investigated optimal combinations of FITs and capital subsidies in the plane where the horizontal and vertical axes represent the FIT and capital subsidy levels, respectively. A line could be drawn in the plane to show the boundary for a household between adopting and not adopting a PV system. In contrast, as will be explained in Sect. 7.2, we consider a plane in which the horizontal and vertical axes are the amount of PV electricity to be generated and the net cost of a PV system, respectively. Accordingly, for a given combination of FITs and capital subsidies, we may draw a line in the plane to separate adopters and non-adopters.

After developing the model, we examine again optimal combinations of FITs and capital subsidies in terms of maximization of PV electricity, minimization of promotion cost, and maximization of social welfare. As will be shown in Sect. 7.3, the same results are obtained as in Chap. 5.

The remainder of this chapter is organized as follows. In Sect. 7.2, we present a new model that is constructed with continuous variables. The definitions of the variables and the decision-making of a household are described. The model in this section supposes that all PV electricity is eligible for FITs. Section 7.3 examines optimal combinations of FITs and capital subsidies. The model is adapted to allow only surplus PV electricity to be eligible for FITs in Sect. 7.4. Section 7.5 concludes the chapter.

7.2 The Model

In this section, we develop a continuous-variable version of the model for investigating a combination of FITs and capital subsidies; first we define variables and then describe the decision-making of a household.

7.2.1 *Definition of Variables*

The situation is the same as that of the residential sector in Chap. 5: a government is seeking to promote solar PV power generation in the residential sector. It offers a combination of FITs and capital subsidies to encourage a certain number of households to adopt PV systems. On the one hand, under the FIT system, an adopter can sell the electricity the PV system generates at a set price during a set period of years. On the other hand, households are granted a capital subsidy as a lump sum payment when they purchase and install a PV system.

The definition of variables is also the same as in Chap. 5 except that each is a continuous variable: z is the amount of PV electricity to be generated during the lifetime of the PV system, v is the satisfaction a household experiences from the adoption of a PV system, and I is the cost of a PV system. Note that these are expressed as a value per PV system, assuming that the PV panels an adopter installs on the rooftop have almost the same capacity.

We set $u = v - I$ because, as will be clear in the modeling below, v and I always appear simultaneously in the form of $v - I$. Let us make an assumption of $u < 0$: this implies, as will also be clear in the modeling below, that no household will adopt a PV system unless FITs and/or capital subsidies are provided by the government.

Let $\hat{q}$ be the quantity of adoption the government is seeking to attain. To this end, it offers a price, p, at which an adopter can sell PV electricity, and a capital subsidy, s, which is granted to an adopter as a lump sum payment at the time of installation. The government is seeking to attain $\hat{q}$ as efficiently as possible in that it sets p and s so that $\hat{q}$ would no longer be attained if p or s decreased infinitesimally.

7.2.2 *Household Decision-Making*

Although the definition of variables is basically the same as in Chap. 5, the most important difference is in the procedure for determining p and s. In Chap. 5, for a household i with values z_i, v_i, and I_i, we first identified the (p, s) in the ps–plane under which household i adopts, and we next worked out the locus of (p, s) on which a certain number of households adopt a PV system. In contrast, in the present model, the zu–plane is considered. Remember $u = v - I$. It is assumed that every household is distributed continuously with respect to (z, u) in the zu–plane, where $z > 0$, $u < 0$. We may then draw a straight line in the zu–plane, which is specified by p and s, and which divides households between adopters and non-adopters. Accordingly, controlling p and s may yield the target quantity of adoption.

The technicalities of determining p, s are as follows. Suppose the government offers a combination of p and s. Then, a household adopts a PV system if and only if $pz + v + s - I \geq 0$, or $pz + s + u \geq 0$. Let $\Omega_{p,s}$ denote the set of (z, u) of the adopters, i.e., $\Omega_{p,s} = \{(z, u) | \ pz + s + u \geq 0, z > 0, u < 0\}$. A downward-sloping line

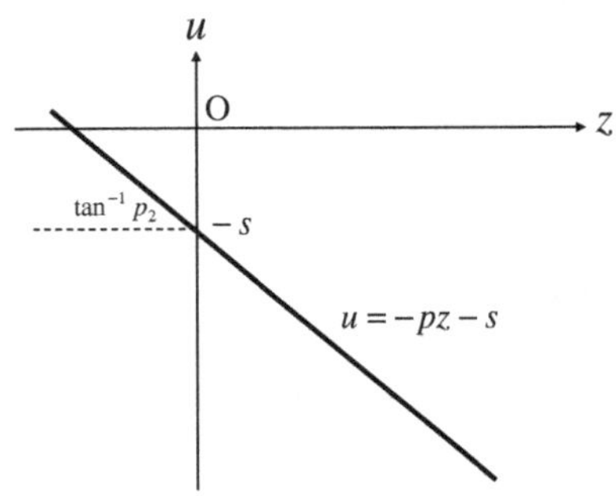

Fig. 7.1 The line dividing households into adopters and non-adopters. The adopters are situated in the area above or on the line, while the non-adopters are below the line, where $z > 0$, $u < 0$

$$u = -pz - s \tag{7.1}$$

forms a boundary in the zu–plane between adopters and non-adopters (Fig. 7.1). The adopters are situated in the area above or on the line, while the non-adopters are below the line.

Given a (p, s), let the quantity of adoption be denoted by $q(p, s)$. From the assumption that households are continuously distributed with respect to (z, u) in the zu–plane, it may be deduced that $q(p, s)$ is a continuous function of p, s. This allows $q(p, s)$ to take any positive value not restricted to an integer. Conversely, for a given $\hat{q}$, either p can be uniquely identified against a fixed s, or s can be uniquely identified against a fixed p, such that $q(p, s) = \hat{q}$.

It is assumed that for a given (z, u), there exists a nonnegative, continuous function, $\pi(z, u)$, whose integral exists on $\Omega_{p,s}$ such that

$$q(p, s) = \iint_{\Omega_{p,s}} \pi(z, u) dz du. \tag{7.2}$$

Intuitively, $\pi(z, u)$ corresponds to the number of households that have z, u. The existence of the integral may be accepted because the total number of households and the values of z and u are bounded in practice.

For a given p, $h(p)$ is defined such that $q(p, h(p)) = \hat{q}$. In other words, for a given p, the s that attains $q(p, s) = \hat{q}$ exists, which we express as $h(p)$. It is assumed that $q(p, 0) < \hat{q}$ so that $h(p) > 0$. Obviously, $h(p)$ is a decreasing function of p. Let $\Omega_p = \Omega_{p,h(p)}$, i.e., $\Omega_p = \{(z, u) |\, u \geq -pz - h(p), z > 0, u < 0\}$. Hence,

$$\hat{q} = \iint_{\Omega_p} \pi(z, u) dz du \tag{7.3}$$

holds.

7.3 Optimal Combinations

We examine optimal combinations of FITs and capital subsidies in terms of maximization of PV electricity, minimization of promotion cost, and maximization of social welfare.

To begin, we show a useful feature of the model. Suppose that p is increased from p_1 to p_2. Then, $h(p_1) > h(p_2)$ follows because $h(p)$ is a decreasing function of p. There exists (z_0, u_0) that uniquely satisfies equations $u_0 = -p_1 z_0 - h(h_1)$ and $u_0 = -p_2 z_0 - h(p_2)$ simultaneously (Fig. 7.2). In other words, (z_0, u_0) is the intersection of the two straight lines, $u = -p_1 z - h(h_1)$ and $u = -p_2 z - h(p_2)$. It is verified that $z_0 > 0$, $u_0 > 0$ from $h(p_1) > h(p_2)$.

7.3.1 *Maximization of PV Electricity*

Let us examine the PV electricity maximization problem:

$$\text{maximize} \iint_{\Omega_p} z\pi(z, u) dz du. \tag{7.4}$$

If p is increased from p_1 to p_2, the change in the PV electricity to be generated is

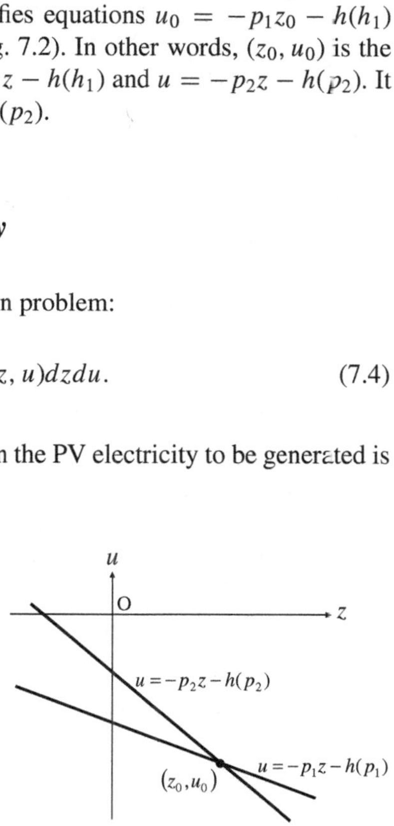

Fig. 7.2 A change in a FIT level. If p is increased from p_1 to p_2, there exists an intersection (z_0, u_0) of the two lines, $u = -p_1 z - h(h_1)$ and $u = -p_2 z - h(p_2)$

$$
\begin{aligned}
&\iint_{\Omega_{p_2}} z\pi(z,u)dzdu - \iint_{\Omega_{p_1}} z\pi(z,u)dzdu \\
&= \iint_{\Omega_{p_2}\backslash\Omega_{p_1}} z\pi(z,u)dzdu - \iint_{\Omega_{p_1}\backslash\Omega_{p_2}} z\pi(z,u)dzdu \\
&> z_C\left(\iint_{\Omega_{p_2}\backslash\Omega_{p_1}} \pi(z,u)dzdu - \iint_{\Omega_{p_1}\backslash\Omega_{p_2}} \pi(z,u)dzdu\right) \\
&= 0. \qquad (7.5)
\end{aligned}
$$

The inequality in (7.5) is obtained from Fig. 7.2: for any $(z, u) \in \Omega_{p_2}\backslash\Omega_{p_1}, z \geq z_0$ holds; for any $(z, u) \in \Omega_{p_1}\backslash\Omega_{p_2}, z \leq z_0$ holds. The last equality results from the fact that the two integrals in the parentheses are equal because the quantity of adoption remains $\hat{q}$. Hence, the amount of PV electricity increases if p_1 is increased to p_2. Therefore, it increases with p. This is the same result obtained from the model with discrete variables in Chap. 5.

7.3.2 *Minimization of Promotion Cost*

Let us examine the promotion cost minimization problem. Recall that the promotion cost consists of the costs of FITs and capital subsidies less the avoided cost of conventional power generation. Thus, the problem is

$$\text{minimize} \iint_{\Omega_p} [(p-c)z + h(p)]\pi(z,u)dzdu. \qquad (7.6)$$

The promotion cost may be reduced as

$$\iint_{\Omega_p} [(p-c)z + h(p)]\pi(z,u)dzdu = \hat{q} \cdot [(p-c)\bar{z} + h(p)], \qquad (7.7)$$

where $\bar{z}$ denotes the average amount of PV electricity over all of the adopters.

Whether $(p - c)\bar{z} + h(p)$ is increasing, decreasing, or invariable is not clear, as shown in the following. Due to the result of Sect. 7.3.1, when p is increased from p_1 to p_2, i.e., $p_1 < p_2$, $\bar{z}$ increases from $\bar{z}_1$ to $\bar{z}_2$, i.e., $\bar{z}_1 < \bar{z}_2$. On the one hand, if $\bar{z}_1, \bar{z}_2$ is sufficiently small, $-p_1\bar{z}_1 - h(p_1) < -p_2\bar{z}_2 - h(p_2)$, or $p_1\bar{z}_1 + h(p_1) > p_2\bar{z}_2 + h(p_2)$ holds (Fig. 7.3a). Hence, $(p_1 - c)\bar{z}_1 + h(p_1) > (p_2 - c)\bar{z}_2 + h(p_2)$. On the other hand, if $\bar{z}_1, \bar{z}_2$ is sufficiently large, $-p_1\bar{z}_1 - h(p_1) > -p_2\bar{z}_2 - h(p_2)$, or $p_1\bar{z}_1 + h(p_1) < p_2\bar{z}_2 + h(p_2)$ holds (Fig. 7.2b). If $p_2 - p_1$ is small enough, so is $\bar{z}_2 - \bar{z}_1$. Accordingly, $(p_1 - c)\bar{z}_1 + h(p_1) < (p_2 - c)\bar{z}_2 + h(p_2)$ may hold. It is thus not definite whether $(p - c)\bar{z}_1 + h(p)$ is increasing, decreasing, or invariable. Therefore, it is unclear whether the promotion cost is increasing, decreasing, or invariable as p

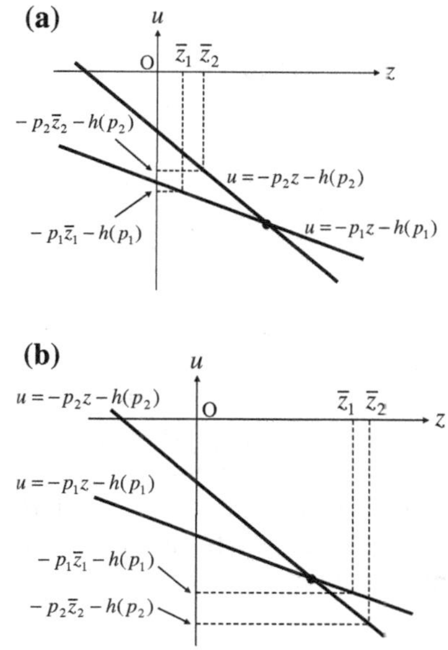

Fig. 7.3 A change in promotion cost associated with the increase in p. Suppose that p is increased from p_1 to p_2, **a** if $\bar{z}_1, \bar{z}_2$ are sufficiently small, the promotion cost will decrease, **b** if $\bar{z}_1, \bar{z}_2$ are sufficiently large, it may increase

increases. The change depends on $\pi(z, u)$. This is the same result obtained by the model with discrete variables in Chap. 5.

7.3.3 *Maximization of Social Welfare*

Let us examine the social welfare maximization problem. Recall that social welfare is the sum of valuation on PV generation v, societal benefit from PV electricity bz, and avoided cost of conventional power generation cz minus the cost I, over all adopters. Taking account of $u = v - I$, the social welfare maximization problem is

$$\text{maximize} \iint_{\Omega_p} [(b+c)z + u]\pi(z, u)dzdu. \tag{7.8}$$

It is shown in the following that social welfare is maximized at $p = b + c$. First, suppose $p_1 < p_2 \leq b+c$. Then, Fig. 7.2 reveals that for $(z, u) \in \Omega_{p_2} \backslash \Omega_{p_1}$, $(b+c)z+u$ is minimized at (z_0, u_0); for $(z, u) \in \Omega_{p_1} \backslash \Omega_{p_2}$, $(b+c)z+u$ is maximized at (z_0, u_0). Accordingly,

$$
\begin{aligned}
&\iint_{\Omega_{p_2}} [(b+c)z+u]\pi(z,u)dzdu - \iint_{\Omega_{p_1}} [(b+c)z+u]\pi(z,u)dzdu \\
&= \iint_{\Omega_{p_2}\backslash\Omega_{p_1}} [(b+c)z+u]\pi(z,u)dzdu - \iint_{\Omega_{p_1}\backslash\Omega_{p_2}} [(b+c)z+u]\pi(z,u)dzdu \\
&\geq [(b+c)z_0+u_0]\left(\iint_{\Omega_{p_2}\backslash\Omega_{p_1}} \pi(z,u)dzdu - \iint_{\Omega_{p_1}\backslash\Omega_{p_2}} \pi(z,u)dzdu\right) \\
&= 0.
\end{aligned}
\tag{7.9}
$$

Hence, the social welfare increases if p is increased from p_1 to p_2.

Next, suppose $b+c \leq p_1 < p_2$. In this case, it should be noted from Fig. 7.2 that for $(z,u) \in \Omega_{p_1}\backslash\Omega_{p_2}$, $(b+c)z+u$ is maximized at (z_0, u_0); for $(z,u) \in \Omega_{p_1}\backslash\Omega_{p_2}$, $(b+c)z+u$ is minimized at (z_0, u_0). As in the first case, it is verified that social welfare decreases if p is increased from p_1 to p_2. Thus, social welfare is maximized at $p = b+c$, which leads to $s = h(b+c)$. This is the same result obtained by the model with discrete variables in Chap. 5.

7.4 Feed-in Tariffs for Surplus PV Electricity

In this section, suppose that only surplus PV electricity, the PV electricity generated that is not self-consumed and is fed into the power grid, is eligible for FITs. We only present the model adapted to this case; determining optimal ps is left as a future research topic.

As in the model in Chap. 5, a household's amount of PV electricity z is divided into surplus PV electricity x and the amount of self-consumed electricity y, where $x+y=z$.

Let r denote the electricity rate that includes the surcharge to finance FITs. Note that if $p < c$, then $r < c$. In particular, if $p = c$, then $r = c$.

Given a combination of p, r and s, a household adopts a PV system if and only if $px+ry+s+u \geq 0$. Let $\Theta_{p,r,s}$ denote the set of (x, y, u) of the adopters, i.e., $\Theta_{p,r,s} = \{(x,y,u) | px+ry+s+u \geq 0\}$. A plane

$$
px+ry+s+u=0 \tag{7.10}
$$

forms a boundary in the xyu–space between adopters and non-adopters. The adopters are situated in the area above or on the plane, while the non-adopters are below the plane.

Given $(p,r,s) \in R_+^3$, let $q(p,r,s)$ denote the quantity of adoption. It is assumed that there is a nonnegative, continuous function, φ, defined in R_+^3 such that the integral of $\varphi(x,y,u)$ on $\Theta_{p,r,s}$ exists, satisfying

$$q(p, r, s) = \iiint_{\Theta_{p,r,s}} \varphi(x, y, u)dxdydu. \tag{7.11}$$

There are two constraints on (p, r, s). First, because the government is seeking to achieve the quantity of adoption $\hat{q}$, it must hold

$$\hat{q} = q(p, r, s). \tag{7.12}$$

Second, the costs of FITs must be equal to the surcharge on the electricity. Let L be the fixed sum of electricity consumption by all households (adopters plus non-adopters). As the model addresses the earlier stage of diffusion of PV systems, it is assumed that L is sufficiently large. The electric utility collects the funds for FITs by adding the extra rate $r - c$ to the standard electricity rate c:

$$(r - c)\left(L - \iiint_{\Theta_{p,r,s}} y\varphi(x, y, u)dxdydu\right) = (p - c)\iiint_{\Theta_{p,r,s}} x\varphi(x, y, u)dxdydu. \tag{7.13}$$

Thus, for a given p, s is specified as a function of p, r, i.e., $s(p, r)$ from Eq. (7.12) together with Eq. (7.11). Then, from Eq. (7.13), r is a function of p, i.e., $r(p)$. Define Θ_p as $\Theta_p = \Theta_{p,r(p),s(p,r(p))}$.

Then, it might be possible to calculate an optimal combination of FITs and capital subsidies that satisfies an optimality criterion. This section briefly describes an approach to investigating the social welfare maximization problem.

This approach is similar to the idea presented in Chap. 5. First, $r(p) \approx c$ may be deduced from Eq. (7.13) because L is sufficiently large. Then, examine how social welfare varies, setting $r(p) = c$ as an approximation. In doing so, $u = -px - cy - s$ from Eq. (7.10) allows us to examine the problem in the *xu*–plane in place of the *zu*–plane in Sect. 7.3.3.

7.5 Conclusion

This chapter provided a version of the model developed in Chap. 5. The present model used continuous variables, rather than discrete variables, to characterize a household that is a potential adopter of a PV system. Then, optimal combinations of FITs and capital subsidies were investigated in terms of maximization of PV electricity, minimization of promotion cost, and maximization of social welfare, yielding the same results as those in Chap. 5.

Important attributes of the investigation in this chapter are the characterization of a household, i.e., using continuous variables instead of discrete variables, and the investigation of optimal combinations of FITs and capital subsidies. In Chap. 5, given a set of variables that characterize a household, we examined what levels of

FITs and capital subsidies enable that household to adopt a PV system. In contrast, in this chapter, given a combination of FITs and capital subsidies, we examine which households adopt a PV system. In reality, it is more likely that households follow a certain distribution, typically a normal distribution, with respect to the associated relevant variables, such as amount of PV electricity to be generated and cost of a PV system. In this regard, the approach in this chapter might be more practical than that in Chap. 5 for deducing policy implications.

The approach to modeling in this chapter is so intuitive that mathematical strictness is set aside to some extent. For example, the existence of a function representing the distribution of households and the computability of the integral of the function are not addressed in a rigorous way. The assumption that the relevant variables associated with a household distribute continuously should be examined empirically. These are important research topics that should be studied in the future.

Part II
Economic and Policy Issues of Renewable Energy

Chapter 8
Promoting the Development of Renewable Energy Under Uncertainty

Abstract Foreign direct investment in renewable energy projects, particularly where biomass is used as input, has been attracting increasing attention, partly due to the Clean Development Mechanism (CDM) of the Kyoto Protocol. In these situations, there may be an information gap between a host country's government and the foreign firm that will invest: while the firm can collect information regarding the project, for example through a feasibility study, it will be difficult for the government to know whether the foreign firm is undertaking the project efficiently. It is supposed that the government will offer the foreign firm some remunerations, consisting of feed-in premiums (FIPs) and capital subsidies, to encourage investment in such a project. The purpose of this chapter is to determine an optimal combination of FIPs and capital subsidies that encourages investment in a CDM project by a foreign firm, while minimizing the cost of FIPs and capital subsidies. To this end, we develop a microeconomic model that accounts for this information gap. It is shown that a single combination of remunerations is determined to be optimal regardless of the existence of such an information gap. The model developed in this chapter may be considered an extension of the model developed in Chap. 5, this time applied to an investigation that accounts for uncertainty.

Keywords Clean development mechanism · Feed-in premium · Capital subsidy
Asymmetric information

8.1 Introduction

So far, we have investigated feed-in tariffs by considering solar and wind as renewable energy sources. However, biomass is increasingly attracting attention as well. One of the driving forces behind the use of biomass will be the Clean Development Mechanism (CDM) of the Kyoto Protocol under the United Nations Framework Convention on Climate Change (UNFCCC).

The CDM was established to assist Annex I Parties to the UNFCCC, mainly developed nations, in achieving compliance with their quantified greenhouse gas emission limitation and reduction commitments under the Kyoto Protocol. If the emission reductions of an approved project are certified, a certificate called Certified

Y. Yamamoto, *Feed-in Tariffs and the Economics of Renewable Energy*,
https://doi.org/10.1007/978-3-319-76864-9_8

Emission Reductions (CERs) is issued; this certificate may be used for complying with commitments, or it may be traded as an economic good. The approved project for the CDM, called the CDM project in this chapter, is also favorable to non-Annex I Parties, mainly developing nations, because it can help them to achieve sustainable development. According to the website of the CDM, UNFCCC, 7789 projects have been registered so far (UNFCCC 2017).

In looking at CDM projects, we can identify a variety of features (for in-depth information on each of the CDM projects, see the website of the CDM, UNFCCC). To study policy instruments designed to promote the use of biomass through the CDM, we consider the following CDM projects: a firm in a developed country is seeking to invest in a CDM project located in a developing country, called a host country, to obtain CERs. The biomass used as input may include bagasse, empty fruit bunches, and rice husks, which exist plentifully in Southeast Asia, for example. The firm builds a power plant at which electricity is generated by burning biomass. This is a typical project, and it has two important features.

First, there may be an information gap between the government of the host country and the firm that is considering investing in a CDM project there. The firm will experience uncertainty about the project because it is investing in a foreign country rather than its own country; these uncertainties concern, for example, the availability of biomass, the suitability of the planned equipment for the biomass in the project, the labor market, and the legal conditions. However, these uncertainties will be resolved, to some extent, by a feasibility study before the firm makes an investment decision. On the other hand, the government of the host country will not have enough information about the foreign firm; it will be difficult to know whether the firm can undertake the project efficiently given the circumstances of the host country. Therefore, it will be the host government, rather than the foreign firm, that must address informational inferiority.

Second, the government of the host country will offer financial support to the firm that invests in a CDM project. The firm must conduct a feasibility study, at some cost, before making an investment decision, and incur investment costs if actually investing. These costs are an obstacle to investment and make the firm reluctant to invest. On the other hand, the government of the host country wants to call for investment from foreign countries to achieve sustainable development, particularly to generate electricity to meet increasing demand. Accordingly, the government of the host country will offer remunerations for a CDM project. These include tax exemption, a favorable price on the electricity generated, capital subsidies, and other benefits.

Taking the two features together, the government must offer remunerations despite its uncertainty regarding whether the foreign firm considering investing can efficiently undertake a project. The purpose of this chapter is to clarify what levels of remunerations the host country's government should offer, when facing an information gap, to encourage a foreign firm to invest in a CDM project.

Specifically, it is supposed that the government offers a combination of feed-in premiums (FIPs) and capital subsidies, where FIPs are a premium added to the wholesale price of electricity and capital subsidies are paid as a lump sum payment. Hence, the government's problem is to determine the optimal levels of FIPs and capital subsidies to minimize their cost. On the one hand, if these levels are too low, the project will not be implemented; on the other hand, if they are too high, excess burdens are imposed on the people of the country because FIPs and capital subsidies are financed through increased electricity rates and taxes.

In Chap. 5, we developed a microeconomic model to investigate an optimal combination of feed-in tariffs and capital subsidies for solar photovoltaic (PV) power generation for the residential and business sectors. In that model, we did not consider the uncertainty that would be involved in determining FIT and capital subsidy levels. The reason is that for electricity generation through PV systems, there may be little uncertainty about the relevant variables. For example, the investment cost, which consists mainly of the cost of equipment, is easier to estimate, and the expected amount of electricity to be generated is projected relatively accurately because there is abundant meteorological data. The operating cost may have little importance in PV generation. Accordingly, to investigate an optimal combination of FIPs and capital subsidies in a CDM project, we must develop another type of model that includes uncertainty.

Models used to investigate technology transfer may help to develop our new model. In those models, it is assumed that there is an information gap between a licensor and a licensee (for example, Gallini and Wright 1990; Das 1999; Marjit and Mukherjee 2001; Choi 2001). Gallini and Wright (1990) investigated licensing contracts in terms of a licensor's profit-maximizing combinations of fixed fees versus output-based royalties. They showed, under the conditions that the licensor has private information about the technology to be licensed and that sharing the information may facilitate imitation, that the licensor signals the type of the technology by using an output-based payment and may leave some of the rents with the licensee. Under the same conditions as those in the model by Gallini and Wright (1990), Marjit and Mukherjee (2001) considered the possibility of equity participation when licensing the technology. They showed that when the buyer of the technology is not certain about the quality of the technology, equity participation may be an effective means of striking a deal for the best available technology.

A similar but more standard study is that by Macho-Stadler and Pérez-Castrillo (2001, pp. 149–153). They examined a contract for licensing a technology, developing a model where an owner of a technology license makes a contract with a firm producing a good by using that technology. The seller offers a pair of payments consisting of a fixed transfer payment and a payment proportional to the production levels. It is supposed that there is an information gap between the seller and the buyer of the license. For example, the buyer knows more than the seller about the market demand for the product or how to best apply the technology to the production process. In this situation, the optimal contract for the seller consists of a menu of contracts: two pairs of fixed and proportional payments.

These studies were concerned with the problem of adverse selection. In economics literature, adverse selection is a type of information gap between two parties: one party, called a principal, makes a contract with another party, called an agent, in order to have a task completed. Before signing the contract, the agent has some private information about the task, which the principal does not know. In this situation, an efficient design of the contract for the principal is to offer several alternative contracts; the agent chooses among these alternatives.

The model to be developed in this chapter follows a model of adverse selection. The outline of this model is that a government in a host country offers remunerations to a foreign firm to encourage investment in a CDM project that will generate electricity from biomass in the country. The remunerations consist of FIPs and capital subsidies; the firm can sell the electricity generated in the project at a price equal to a wholesale price plus FIPs and is granted capital subsidies as an up-front lump sum payment. The government wants to minimize the cost of these remunerations. In this sense, the model in this chapter is an extension of the model developed in Chap. 5; the present model includes uncertainty in that the government does not know whether the foreign firm is of good quality.

The remainder of this chapter is organized as follows. In Sect. 8.2, the basic model is developed: it is assumed as a benchmark that there is no information gap between the host country's government and the firm considering investing in a project in the host country. Then, the optimal combination of FIPs and capital subsidies is calculated. In contrast, in Sect. 8.3, it is supposed that there is an information gap between the government and the firm: the government does not know how efficiently the foreign firm will complete the project. Then, the optimal contract is calculated again. Finally, Sect. 8.4 concludes the chapter.

8.2 The Model

In this section, we develop a model to examine the combination of FIPs and capital subsidies a government should offer to encourage a foreign firm to invest in a project that will generate electricity from biomass in the host country. In this section, it is assumed as a benchmark that there is no information gap between the host country's government and the foreign firm considering investment. In the next section, the model is adapted to account for the information gap and to examine its effect on the optimal combination of FIPs and capital subsidies.

8.2.1 Definition of Variables

The government of a developing country is considering increasing electricity output to meet increasing demand, and it also wants to utilize the country's abundant biomass. To invite investment in this project from foreign countries, the government offers a foreign firm two types of remunerations, FIPs and capital subsidies.

Suppose that a firm is considering investing in such a project. Let F be the cost of the investment in the project, which may be considered as the fixed cost of the project.

Let x be the amount of electricity generated in the project, designated as green electricity. Let $C(x)$ denote the cost of green electricity generation. It is assumed that $C'(x) > 0,\ C''(x) > 0$, reflecting the fact that as the amount of green electricity increases, the amount of biomass to be input must increase such that collecting additional biomass becomes costly. It is also assumed that $C(0) = 0$. In other words, the total cost is $C(x) + F$, where the variable cost is $C(x)$ and the fixed cost is F.

The government of the host country offers FIPs and capital subsidies. Let s be a capital subsidy; the firm can be granted a lump sum payment s at the outset of the project. On the other hand, let p be the level of an FIP; the firm can sell a unit of green electricity at the electricity wholesale price plus p.

The firm can obtain a certificate that proves the amount of green electricity generated, which may be sold at a price on an international certificate market. It is assumed that the certificate market is competitive such that a fixed price is given for every market participant. At the same time, the firm can sell the green electricity at the wholesale electricity price plus p. Let r denote the sum of the wholesale electricity price and the certificate price, which is fixed in the model. Accordingly, the firm can obtain $p + r$ per unit of green electricity.

It is assumed that if the government does not provide any remunerations, i.e., $p = s = 0$, the firm cannot make a profit, and thus it does not invest in the project. On the other hand, it is postulated that the host country's government seeks to minimize the cost of FIPs and capital subsidies due to its limited financial resources. Hence, the government faces a trade-off between incentivizing a foreign firm to invest and minimizing the cost of FIPs and capital subsidies.

8.2.2 The Contract Minimizing the Cost

To begin, let us determine the optimal production level of green electricity. Let $\pi(p, x)$ be a producer surplus that the firm makes by producing green electricity x for a given p once it has invested in the project. Specifically,

$$\pi(p, x) = (p + r)x - C(x). \tag{8.1}$$

Let us assume $C'(0) < r$ so that a solution to $\partial\pi/\partial x = p + r - C'(x) = 0$ exists where $x > 0$ even if $p = 0$. This means that once the firm has invested in a project, it can make a profit even if FIPs are not provided by the government. Because $\partial^2\pi/\partial x^2 = -C''(x) < 0$, the x that maximizes $\pi(p, x)$, denoted by $x^*(p)$, satisfies

$$p + r = C'\left(x^*(p)\right). \tag{8.2}$$

Because the optimal production level x^* is a function of p, we hereafter simply write $\pi\ (x^*(p))$ for $\pi\ (p, x^*(p))$. Therefore, for a given combination of p and s, the profit the firm makes from the project is

$$\Pi(p, s) = \pi\left(x^*(p)\right) + s - F. \tag{8.3}$$

Two remarks are in order. First, Eq. (8.2) yields

$$\frac{dx^*(p)}{dp} = \frac{1}{C''(x^*(p))} > 0, \tag{8.4}$$

which means that the optimal production level is an increasing function of p, a natural result. Second, by plugging $x^*(p)$ into Eq. (8.1) and differentiating with respect to p, we obtain

$$\frac{d\pi(x^*(p))}{dp} = x^*(p), \tag{8.5}$$

which means that the profit is an increasing function of p, another natural result.

The firm invests in a project if and only if $\pi(x^*(p)) + s - F \geq 0$. Because we have assumed that the firm would not invest in the project if the government did not offer any remuneration,

$$\pi\left(x^*(0)\right) - F < 0 \tag{8.6}$$

holds.

The government wants to minimize the cost of remunerations, $px + s$, while it also wants a foreign firm to invest in the project. Hence, the problem for the government is

$$\text{minimize} \quad px^*(p) + s, \tag{8.7}$$

$$\text{subject to} \quad \pi\left(x^*(p)\right) + s - F \geq 0, \tag{8.8}$$

$$p \geq 0, \tag{8.9}$$

$$\text{and} \quad s \geq 0. \tag{8.10}$$

This may be solved as follows. The Lagrangian L is

$$L = px^*(p) + s - \lambda_1\left[\pi\left(x^*(p)\right) + s - F\right] - \lambda_2 p - \lambda_3 s, \tag{8.11}$$

where λ_1, λ_2, and λ_3 are Lagrange multipliers.

We now write out the complete set of first order conditions:

$$\frac{\partial L}{\partial p} = x^*(p) + \frac{p}{C''(x^*(p))} - \lambda_1 x^*(p) - \lambda_2 = 0, \tag{8.12}$$

$$\frac{\partial L}{\partial s} = 1 - \lambda_1 - \lambda_3 = 0, \tag{8.13}$$

$$\lambda_1\left[\pi\left(x^*(p)\right) + s - F\right] = 0, \tag{8.14}$$

$$\lambda_2 p = 0, \tag{8.15}$$

$$\lambda_3 s = 0, \tag{8.16}$$

$$\pi\left(x^*(p)\right) + s - F \geq 0, \tag{8.17}$$

$$p \geq 0,\ s \geq 0, \tag{8.18}$$

$$\text{and}\ \ \lambda_1, \lambda_2, \lambda_3 \geq 0. \tag{8.19}$$

We used Eqs. (8.4) and (8.5) to derive Eq. (8.12).

First, suppose $\lambda_3 > 0$. From Eq. (8.16), $s = 0$. Then, from condition (8.6), $p > 0$, which leads to $\lambda_2 = 0$ due to Eq. (8.15). Because $\lambda_1 = 1 - \lambda_3 < 1$ from Eq. (8.13), the left hand side of Eq. (8.12) is

$$(1 - \lambda_1)\, x^*(p) + \frac{p}{C''\left(x^*(p)\right)} > 0, \tag{8.20}$$

which contradicts Eq. (8.12). Therefore, $\lambda_3 = 0$.

Accordingly, $\lambda_1 = 1$ from Eq. (8.13). Then, from Eq. (8.12),

$$\lambda_2 = \frac{p}{C''\left(x^*(p)\right)}. \tag{8.21}$$

Plugging Eq. (8.21) into Eq. (8.15) yields $p = 0$. Then, because $\lambda_1 = 1$, $s = F - \pi\left(x^*(0)\right)$ from Eq. (8.14).

Therefore, the solution to the minimization problem is

$$p^* = 0,\ s^* = F - \pi\left(x^*(0)\right). \tag{8.22}$$

In other words, the government should set the FIP level to zero, while the capital subsidy level is set so that the profit to the firm equals zero.

The result may be graphically illustrated (Fig. 8.1). From Eqs. (8.4) and (8.5), the firm's profit, $\pi\left(x^*(p)\right) + s - F$, becomes large as p or s increases. In other words, the iso-profit curve located at the upper right in the ps–plane corresponds to a higher level of profit. The slope of the iso-profit curve is

$$\left.\frac{ds}{dp}\right|_{\text{Firm}} = -x^*(p). \tag{8.23}$$

On the other hand, the cost to the government, $px^*(p) + s$, becomes small as p or s decreases. In other words, the iso-cost curve located at the lower left in the ps–plane corresponds to a lower level of cost. The slope of the iso-cost curve is

$$\left.\frac{ds}{dp}\right|_{\text{Government}} = -x^*(p) - \frac{p}{C''\left(x^*(p)\right)}. \tag{8.24}$$

The comparison of (8.23) with (8.24) indicates that the iso-cost curve of the government is always steeper than the iso-profit curve of the firm in the ps–plane (Fig. 8.1). Hence, the government, which wants the firm to invest in the project

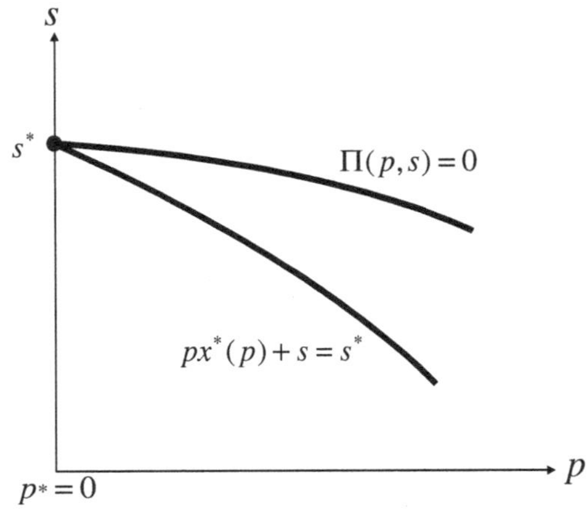

Fig. 8.1 The cost-minimizing combination of FIPs and capital subsidies. The government chooses (p, s) on $\Pi(p, s) = 0$ so that $px^*(p) + s$ is as small as possible

while also minimizing the cost, will choose the (p, s) that minimizes the cost on the iso-profit curve giving null profit to the firm, which is precisely the combination presented in (8.22).

8.3 Asymmetric Information

In Sect. 8.2, we investigated the case of symmetric information, where there is no information gap between the host country government and the firm investing in a project. In reality, however, if the investor is a foreign firm, the host country government may not have much information about the firm, especially regarding whether the firm will undertake the project efficiently. In this section, it is assumed that there may be two types of firms, type G and type B. Type G is more efficient than type B in that the type G firm incurs lower costs than the type B firm in producing green electricity.

Let $C_G(x)$ and $C_B(x)$ denote the cost of producing green electricity x by the firms of type G and type B, respectively. Assume that $C'_G(x) < C'_B(x)$ because type G is more efficient than type B. In addition, it is assumed that regardless of the type of firm, the investment cost is F and the fixed cost of production is zero: $C_G(0) = C_B(0) = 0$. In other words, it is impossible for the government to distinguish between type G and type B firms simply by observing the investment cost.

The producer surplus to each type is denoted by $\pi_G\left(x^*_G(p)\right) = (p + r)x^*_G(p) - C_G\left(x^*_G(p)\right)$ for type G and $\pi_B\left(x^*_B(p)\right) = (p + r)x^*_B(p) - C_B\left(x^*_B(p)\right)$ for type B. Accordingly, the profit is $\Pi_G(p, s) = \pi_G\left(x^*_G(p)\right) + s - F$ for type G and $\Pi_B(p, s) = \pi_B\left(x^*_B(p)\right) + s - F$ for type B.

The only information that the government knows regarding the type of the firm is the probability that the firm is of a particular type: the firm is of type G with probability t and of type B with probability $(1-t)$. Accordingly, the government offers two combinations of FIPs and capital subsidies: (p_G, s_G) and (p_B, s_B). The rationale behind these offers is as follows. The government aims to increase electricity generation in the country and wants to make a contract with the firm regardless of its type. At the same time, the government wants to minimize the cost of remunerations. Thus, the government designs two potential contracts so that the firm chooses (p_G, s_G) if it is of type G; otherwise, it chooses (p_B, s_B).

Hence, the problem of the government is

$$\text{minimize} \;\; t\left(p_G x_G^*\left(p_G\right)+s_G\right)+(1-t)\left(p_B x_B^*\left(p_B\right)+s_B\right), \tag{8.25}$$

$$\text{subject to} \;\; \pi_G\left(x_G^*\left(p_G\right)\right)+s_G-F \geq \pi_G\left(x_G^*\left(p_B\right)\right)+s_B-F, \tag{8.26}$$

$$\pi_B\left(x_B^*\left(p_B\right)\right)+s_B-F \geq \pi_B\left(x_B^*\left(p_G\right)\right)+s_G-F, \tag{8.27}$$

$$\pi_G\left(x_G^*\left(p_G\right)\right)+s_G-F \geq 0, \tag{8.28}$$

$$\pi_B\left(x_B^*\left(p_B\right)\right)+s_B-F \geq 0, \tag{8.29}$$

$$p_G \geq 0, \; p_B \geq 0, \tag{8.30}$$

$$\text{and} \;\; s_G \geq 0, \; s_B \geq 0, \tag{8.31}$$

where $x_G^*(p)$ and $x_B^*(p)$ are, for a given p, the optimal levels of green electricity production of the firm of type G and that of type B, respectively: $p+r = C_G'\left(x_G^*(p)\right), \; p+r = C_B'\left(x_B^*(p)\right)$. Inequalities (8.26) and (8.27) are called incentive compatibility constraints because they regulate a firm so that it correctly chooses a combination of remunerations designed for its type. On the other hand, inequalities (8.28) and (8.29) are called participation constraints because they dictate that a firm enters into a contract irrespective of the type.

As shown in Appendix 1 the solution to this minimization problem is

$$p_G = p_B = 0, \; s_G = s_B = F - \pi_B\left(x_B^*(0)\right). \tag{8.32}$$

The government should offer the same levels of FIPs and capital subsidies for both types: while the FIPs are set at zero, the capital subsidies are set so that the profit to the type B firm equals zero. In other words, the government's offer in the asymmetric information case is a singleton, which equals the offer to the type B firm in the symmetric information case. Hence, if the firm is type G, it can make a strictly positive profit.

The result may be graphically illustrated (Fig. 8.2). For any amount of profit, the iso-profit curve of type B is located strictly above the iso-profit curve of type G in the ps–plane because, for a given (p, s), $\Pi_G(p, s) = \pi_G\left(x_G^*(p)\right)+s-F > \pi_B\left(x_B^*(p)\right)+s-F = \Pi_B(p, s)$. In particular, $\Pi_B(p, s) = 0$ is located strictly above $\Pi_G(p, s) = 0$ in the ps–plane. Hence, if the government were to set (p_G, s_G) so that $\Pi_G(p, s) = 0$ for type G, and (p_B, s_B) so that $\Pi_B(p, s) = 0$ for type B, the firm would always choose the (p_B, s_B) regardless of its type. Accordingly, if the

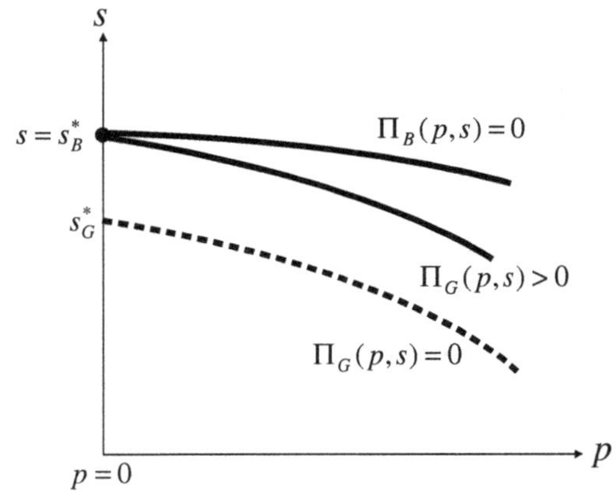

Fig. 8.2 The combination of FIPs and capital subsidies in a case of asymmetric information. The government chooses (p, s) on $\Pi_B(p, s) = 0$. Then, the firm makes a positive profit if it is of type G

government wants the firm to invest in the project regardless of the firm's type, the firm will make a positive profit if it is of type G.

8.4 Conclusion

In this chapter, we investigated foreign direct investment in a CDM project in which biomass is used as input. A government in the host country offered FIPs and capital subsidies to call for investment from a foreign firm while minimizing the costs of FIPs and capital subsidies. We calculated optimal combinations of FIPs and capital subsidies when the government either does or does not know the type of foreign firm that will invest in the project.

It is found that first, if the government knows the type of the firm, the optimal combination of FIPs and capital subsidies is such that the FIP level is set at zero, while the capital subsidy is set so that the firm's profit equals zero. Second, if the government does not have certain information about the type of the firm, it should also offer a single combination of FITs and capital subsidies: the FIP level should be set at zero, and the capital subsidy should be set so that the profit to the firm of a less efficient type equals zero; the firm of a more efficient type can make a positive profit.

The model developed in this chapter may be considered a variation on the model developed in Chap. 5. In that model, any uncertainty was not taken into account. On the other hand, in the present model, it is assumed that the government has uncertain information of the type of the firm, which is a single potential investor in a renewable energy project.

The result that the government should offer a single combination of FIPs and capital subsidies is different from the result of a standard economic analysis of a situation where there is an information gap between a principal and an agent in a contract. In a standard analysis, it is optimal to offer a menu of combinations rather than a single combination. This may be attributable to the shapes of the iso-profit curves.

It is assumed that the government wants to minimize the cost of FIPs and capital subsidies. If another objective function is assumed, the result would have a different form. For example, we may set an objective function as containing benefits yielded by electricity generation in the country. This is an important avenue for future research.

This chapter is concerned with the use of biomass to generate electricity in a developing country. However, regarding the use of biomass, there are other types of projects. For example, wood residues from sawmills are burnt to produce heat and electricity. Agricultural residues and food waste are utilized to produce compost. These projects are strongly connected with the area where the biomass is supplied. Accordingly, the role played by the local government is important if such a project is to be successful. For example, a municipality should organize such a project and promote participation from relevant parties. This topic will be discussed in Chap. 11.

Appendix 1: The Solution to the Minimization Problem in the Case of Asymmetric Information

To begin, note that for a firm of type G, the participation constraint (8.28) may be eliminated because

$$\begin{aligned} \Pi_G\left(x_G^*\left(p_G\right)\right)+s_G-F &\geq \pi_G\left(x_G^*\left(p_B\right)\right)+s_B-F \\ &> \pi_B\left(x_B^*\left(p_B\right)\right)+s_B-F. \\ &\geq 0 \end{aligned} \tag{8.33}$$

The first inequality in (8.33) follows from (8.26), and the third inequality follows from (8.29). The second inequality follows from $C_G'(x) < C_B'(x)$, $p+r = C_G'\left(x_G^*(p)\right)$, and $p+r = C_B'\left(x_B^*(p)\right)$.

The Lagrangian $\tilde{L}$ is

$$\begin{aligned} \tilde{L} =& t\left(p_G x_G^*\left(p_G\right)+s_G\right)+(1-t)\left(p_B x_B^*\left(p_B\right)+s_B\right) \\ &-\mu_1\left[\left(\pi_G\left(x_G^*\left(p_G\right)\right)+s_G-F\right)-\left(\pi_G\left(x_G^*\left(p_B\right)\right)+s_B-F\right)\right] \\ &-\mu_2\left[\left(\pi_B\left(x_B^*\left(p_B\right)\right)+s_B-F\right)-\left(\pi_B\left(x_B^*\left(p_G\right)\right)+s_G-F\right)\right], \\ &-\mu_3\left[\pi_B\left(x_B^*\left(p_B\right)\right)+s_B-F\right] \\ &-\mu_4 p_G-\mu_5 p_B-\mu_6 s_G-\mu_7 s_B \end{aligned} \tag{8.34}$$

where μ_i $(i=1,\ldots,7)$ is a Lagrange multiplier.

Then, we write out the complete set of first order conditions:

$$\frac{\partial \widetilde{L}}{\partial p_G} = tx_G^*(p_G) + \frac{tp_G}{C_G''\left(x_G^*(p_G)\right)} - \mu_1 x_G^*(p_G) + \mu_2 x_B^*(p_G) - \mu_4 = 0, \quad (8.35)$$

$$\frac{\partial \tilde{L}}{\partial p_B} = (1-t)x_B^*(p_B) + \frac{(1-t)p_B}{C_B''\left(x_B^*(p_B)\right)} + \mu_1 x_G^*(p_B)$$

$$- \mu_2 x_B^*(p_B) - \mu_3 x_B^*(p_B) - \mu_5 = 0, \quad (8.36)$$

$$\frac{\partial \widetilde{L}}{\partial s_G} = t - \mu_1 + \mu_2 - \mu_6 = 0, \quad (8.37)$$

$$\frac{\partial \widetilde{L}}{\partial s_B} = (1-t) + \mu_1 - \mu_2 - \mu_3 - \mu_7 = 0, \quad (8.38)$$

$$\mu_1\left[\left\{\pi_G\left(x_G^*(p_G)\right) + s_G - F\right\} - \left\{\pi_G\left(x_G^*(p_B)\right) + s_B - F\right\}\right] = 0, \quad (8.39)$$

$$\mu_2\left[\left\{\pi_B\left(x_B^*(p_B)\right) + s_B - F\right\} - \left\{\pi_B\left(x_B^*(p_G)\right) + s_G - F\right\}\right] = 0, \quad (8.40)$$

$$\mu_3\left\{\pi_B\left(x_B^*(p_B)\right) + s_B - F\right\} = 0, \quad (8.41)$$

$$\mu_4 p_G = 0, \quad (8.42)$$

$$\mu_5 p_B = 0, \quad (8.43)$$

$$\mu_6 s_G = 0, \quad (8.44)$$

$$\mu_7 s_B = 0, \quad (8.45)$$

$$\left\{\pi_G\left(x_G^*(p_G)\right) + s_G - F\right\} - \left\{\pi_G\left(x_G^*(p_B)\right) + s_B - F\right\} \geq 0, \quad (8.46)$$

$$\left\{\pi_B\left(x_B^*(p_B)\right) + s_B - F\right\} - \left\{\pi_B\left(x_B^*(p_G)\right) + s_G - F\right\} \geq 0, \quad (8.47)$$

$$\pi_B\left(x_B^*(p_B)\right) + s_B - F \geq 0, \quad (8.48)$$

$$p_G \geq 0,\ p_B \geq 0,\ s_G \geq 0,\ s_B \geq 0, \quad (8.49)$$

$$\text{and} \quad \mu_1, \mu_2, \mu_3, \mu_4, \mu_5, \mu_6, \mu_7 \geq 0. \quad (8.50)$$

First, suppose $\mu_1 = 0$. From Eq. (8.37), $\mu_6 = t + \mu_2 > 0$. Thus, $s_G = 0$ from Eq. (8.44). On the other hand, $\mu_4 > 0$ from Eq. (8.35). Hence, $p_G = 0$ from Eq. (8.42). However, $p_G = 0$, $s_G = 0$ does not satisfy (8.33). Therefore, $\mu_1 > 0$.

Next, we will show $p_G = p_B$. If $p_G < p_B$, $p_B > 0$ follows from (8.49). Thus, $\mu_5 = 0$ from Eq. (8.43). Because $\mu_2 + \mu_3 = (1-t) + \mu_1 - \mu_7$ from Eq. (8.38), the left hand side of Eq. (8.36) is

$$\frac{\partial \tilde{L}}{\partial p_B} = (1-t)x_B^*(p_B) + \frac{(1-t)p_B}{C_B''\left(x_B^*(p_B)\right)} + \mu_1 x_G^*(p_B) - \left[(1-t) + \mu_1 - \mu_7\right] x_B^*(p_B)$$

$$= \frac{(1-t)p_B}{C_B''\left(x_B^*(p_B)\right)} + \mu_1\left(x_G^*(p_B) - x_B^*(p_B)\right) + \mu_7 x_B^*(p_B)$$

$$> 0. \quad (8.51)$$

However, this contradicts Eq. (8.36). Hence, $p_G \geq p_B$.

If $p_G > p_B$, $p_G > 0$ follows from (8.49). Then, $\mu_4 = 0$ from Eq. (8.42). Furthermore, $\mu_2 = 0$: if $\mu_2 > 0$,

$$\pi_B\left(x_B^*\left(p_B\right)\right)+s_B=\pi_B\left(x_B^*\left(p_G\right)\right)+s_G \tag{8.52}$$

holds from Eq. (8.40). On the other hand, because $\mu_1>0$,

$$\pi_G\left(x_G^*\left(p_G\right)\right)+s_G=\pi_G\left(x_G^*\left(p_B\right)\right)+s_B \tag{8.53}$$

holds from Eq. (8.39). Taken together, Eqs. (8.52) and (8.53) yield

$$\pi_G\left(x_G^*\left(p_G\right)\right)-\pi_B\left(x_B^*\left(p_G\right)\right)=\pi_G\left(x_G^*\left(p_B\right)\right)-\pi_B\left(x_B^*\left(p_B\right)\right). \tag{8.54}$$

However, this is not the case: the left hand side must be larger than the right hand side because $p_B > p_B$. Hence, $\mu_2=0$.

Then, because $\mu_1=t-\mu_6$ from Eq. (8.37), the left hand side of Eq. (8.35) is

$$\begin{aligned}\frac{\partial\tilde{L}}{\partial p_G}&=tx_G^*\left(p_G\right)+\frac{tp_G}{C_G''\left(x_G^*\left(p_G\right)\right)}-\left(t-\mu_6\right)x_G^*\left(p_G\right)\\&=\mu_6x_G^*\left(p_G\right)+\frac{tp_G}{C_G''\left(x_G^*\left(p_G\right)\right)}\\&>0.\end{aligned} \tag{8.55}$$

This contradicts Eq. (8.35). Hence, $p_G=p_B$ holds. Accordingly, $s_G=s_B$ follows from (8.53).

Therefore, the minimization problem is reduced to the following minimization problem with respect to p, s:

$$\text{minimize}\quad t\left(px_G^*(p)+s\right)+(1-t)\left(px_B^*(p)+s\right), \tag{8.56}$$

$$\text{subject to}\quad \pi_B\left(x_B^*(p)\right)+s-F\geq 0, \tag{8.57}$$

$$p\geq 0, \tag{8.58}$$

$$\text{and}\quad s\geq 0. \tag{8.59}$$

The Lagrangian $\widehat{L}$ is

$$\begin{aligned}\widehat{L}=&\,t\left(px_G^*(p)+s\right)+(1-t)\left(px_B^*(p)+s\right)\\&-\xi_1\left[\pi_B\left(x_B^*(p)\right)+s-F\right]-\xi_2p-\xi_3s,\end{aligned} \tag{8.60}$$

where ξ_1, ξ_2, and ξ_3 are Lagrange multipliers.

We now obtain the set of first-order conditions:

$$\frac{\partial\widehat{L}}{\partial p}=tx_G^*(p)+\frac{tp}{C_G''\left(x_G^*(p)\right)}+(1-t)x_B^*(p)+\frac{(1-t)p}{C_B''\left(x_B^*(p)\right)}-\xi_1x_B^*(p)-\xi_2=0, \tag{8.61}$$

$$\frac{\partial\widehat{L}}{\partial s}=1-\xi_1-\xi_3=0, \tag{8.62}$$

$$\xi_1 \left[\pi \left(x_B^*(p)\right) + s - F\right] = 0, \tag{8.63}$$

$$\xi_2 p = 0, \tag{8.64}$$

$$\xi_3 s = 0, \tag{8.65}$$

$$\pi \left(x_B^*(p)\right) + s - F \geq 0, \tag{8.66}$$

$$p \geq 0, \; s \geq 0, \tag{8.67}$$

$$\text{and} \;\; \xi_1, \xi_2, \xi_3 \geq 0. \tag{8.68}$$

First, suppose $\xi_1 = 0$. Then, $\xi_3 = 1 > 0$ from Eq. (8.62). Hence, $s = 0$ from Eq. (8.65). Furthermore, $\xi_2 > 0$ from Eq. (8.61). Thus, $p = 0$ from Eq. (8.64). However, $p = 0$, $s = 0$ does not satisfy (8.66) due to assumption (8.6). Therefore, $\xi_1 > 0$ holds.

Accordingly, from Eq. (8.63),

$$\pi \left(x_B^*(p)\right) + s - F = 0 \tag{8.69}$$

holds.

Next, suppose $\xi_2 = 0$. Then, because $\xi_1 = 1 - \xi_3$ from Eq. (8.62), the left hand side of Eq. (8.61) is

$$\begin{aligned}\frac{\partial \hat{L}}{\partial p} &= t x_G^*(p) - \frac{tp}{C_G''\left(x_G^*(p)\right)} + (1-t)x_B^*(p) + \frac{(1-t)p}{C_B''\left(x_B^*(p)\right)} - (1-\xi_3)\, x_B^*(p) \\ &= t\left(x_G^*(p) - x_B^*(p)\right) + \xi_3 x_B^*(p) + \frac{tp}{C_G''\left(x_G^*(p)\right)} + \frac{(1-t)p}{C_B''\left(x_B^*(p)\right)} \\ &> 0.\end{aligned} \tag{8.70}$$

This contradicts Eq. (8.61). Hence, $\xi_2 > 0$.

Accordingly, $p = 0$ from Eq. (8.64). Then, Eq. (8.69) yields $s = F - \pi\left(x_B^*(0)\right)$.

References

Choi JP (2001) Technology transfer with moral hazard. Int J Ind Organ 19:249–266

Das SP (1999) Direct foreign investment versus licensing. Rev Dev Econ 3:86–97

Gallini NT, Wright BD (1990) Technology transfer under asymmetric information. Rand J Econ 21:147–160

Macho-Stadler I, Pérez-Castrillo D (2001) An introduction to the economics of information. Oxford University Press, Oxford

Marjit S, Mukherjee A (2001) Technology transfer under asymmetric information: the role of equity participation. J Inst Theor Econ 157:282–300

UNFCCC (2017) CDM projects. http://cdm.unfccc.int. Accessed 30 Nov 2017

Chapter 9
Allocation of Ancillary Service Costs to Distributed Generators

Abstract As an increasing amount of the electricity generated from renewable energy sources (RES-E) is fed into the power grid, various problems, such as frequency and voltage instability in the power system, occur more frequently. To address this problem, a system operator provides ancillary services such as balancing electricity supply and demand and procuring reactive power supply. Accordingly, the cost of ancillary services should be appropriately allocated to distributed generators of RES-E. This chapter proposes a method for solving this cost allocation problem. The method proposed is an application of the Aumann–Shapley (A–S) rule, which is one of cost sharing rules among multiple entities. If an ancillary service cost is expressed as a function of an electricity output vector, each element of which corresponds to each distributed generator, the cost share of a distributed generator will be computed based on the A–S rule. The difficulty of this method lies in how to obtain an ancillary service cost function. This chapter proposes that parametric linear programming be used to form that cost function, and we explain this computation method. The method may be useful for designing a new type of feed-in tariff system, which will be necessary after a diffusion goal is achieved under the current feed-in tariff system.

Keywords Distributed generator · Ancillary service · Aumann–shapley rule
Parametric linear programming · Feed-in tariff

9.1 Introduction

An increasing number of renewable energy technologies, such as solar photovoltaic (PV) and wind-power systems, have been deployed mainly as a result of technological developments and various support policies. These systems, which are considered distributed generators, are typically connected to the power grid, into which the electricity generated is fed.

As the amount of electricity generated from renewable energy sources (RES-E) increases, certain difficulties are more likely to arise with regard to the electricity

Y. Yamamoto, *Feed-in Tariffs and the Economics of Renewable Energy*,
https://doi.org/10.1007/978-3-319-76864-9_9

supply. In particular, frequency and voltage instability are two crucial problems with supplying electricity to customers. Frequency stability is equivalent to power balancing in that they are two sides of the same service (Stoft 2002, p. 236). In other words, to keep frequency stable, supply and demand must be balanced. However, the RES-E, especially from solar PV and wind-power generation, will cause unpredictable fluctuations in generation. Furthermore, during some periods of the year, the supply of RES-E may far exceed the demand for electricity.

On the other hand, voltage instability will be caused by reverse power flow. Consider a radial power system where an alternating current (AC) flows along a transmission line from a substation to the end. The voltage gradually drops toward the line end in a normal state. However, if many distributed generators, e.g., solar PV and wind-power generators, are connected at intermediate points on the line, the AC may flow in reverse from the end toward the substation, causing the line voltage to rise and resulting in excess voltage.

These problems must be avoided to maintain proper conditions of the power system in terms of qualities such as stability, reliability, and safety. A system operator plays a role in ensuring a high standard of quality. For example, it calls for spinning reserves starting within 10 minutes in the case of unexpected outages in power generation, which can occur with distributed generators such as solar PV and wind-power systems (Stoft 2002, p. 237). To cope with excess voltage, the system operator adequately secures reactive power: in general, generators produce both real and reactive power. While real power is the normal electrical power traded on the market, reactive power is a necessary part of the transmission of real AC power (Stoft 2002, p. 384). Reactive power can be supplied by capacitors and generators; in the latter case, generators incur an opportunity cost because supplying reactive power reduces the ability to produce real power despite the generator's consuming essentially no fuel (Stoft 2002, p. 238).

The system operator's tasks, such as maintaining frequency and voltage stability, are ancillary services (Stoft 2002, Chap. 3). The cost of providing ancillary services has so far been included in the bundled electricity prices that customers pay (Lin et al. 2006a). However, this cost has been steadily increasing as an increasing amount of RES-E has been fed into the grid. Accordingly, it may make sense to make the grid-connected, distributed generators pay this cost because they are among the beneficiaries of ancillary services: it may not be an exaggeration to say that they can sell RES-E to the grid because of ancillary services.

Thus, the system operator's problem is how to allocate the cost of ancillary services to distributed generators. It might be reasonable to share the cost among all distributed generators on the basis of the amount of electricity they feed into the grid. However, the stress each generator places on the power system may vary disproportionately (Kirschen and Strbac 2004, p. 129).

Numerous studies have addressed this problem. For example, many researchers have been concerned with the cost of reactive power supply, including Choi et al. (1998), Lin et al. (2006a), Ueki et al. (2008), and Koide et al. (2012). They generally computed the reactive power cost to a distributed generator, positing a cost function of real power production by means of an optimization technique, with constraints such as real and reactive power flow equations, power output limits, and voltage limits. These constraints were based on the total real and reactive power output and did not account for the marginal effect of real power output on the reactive power required.

In this chapter, we propose a novel type of rule for sharing the cost of ancillary services. It is supposed that each distributed generator owes a unit of the electricity fed into the grid to a certain amount of ancillary service provided by a system operator in the power system as a whole. Our calculation applies the Aumann–Shapley rule, the exposition of which is the main part of this chapter.

The remainder of the chapter is organized as follows. Section 9.2 provides an exposition of the Aumann–Shapley rule, which will be applied to our problem. In Sect. 9.3, the technicalities of applying the rule to the current problem are explained. Section 9.4 concludes with suggestions for future research.

9.2 The Aumann–Shapley Rule and Its Applications

We propose a rule for allocating the cost of ancillary services. Our proposed rule is based on a well-known cost-sharing rule, the Aumann–Shapley rule. In this section, we first explain the Aumann–Shapley rule and then review prior works that have applied it to the present problem in a different way from our method.

9.2.1 The Aumann–Shapley Rule

We explain one cost-sharing rule, the Aumann–Shapley (A–S) rule (Aumann and Shapley 1974), which we apply to the problem of allocating ancillary service costs. The A–S rule may be seen as an extension of another well-known cost sharing rule, the Shapley value, which is, roughly speaking, an average of an individual's marginal contribution to coalitional outcomes when that individual participates in a coalition. Hence, a good approach to understanding the A–S rule is to consider the Shapley value in the first place. A brief explanation of the rules, at the expense of mathematical rigor, is presented in this section, based on the explanation by Hougaard (2009).

Let us consider a problem of dividing costs among multiple agents. Suppose that there are n agents in an organization N, i.e., $n = |N|$. Let the cost associated with a coalition of agents S, where $|S| = s \leq n$, be given by $c(S)$. The problem is to divide $c(N)$ among the n agents.

The Shapley value considers the order in which an agent joins a coalition and its contribution to the increase in the cost. For a given coalition S, if an agent $i \notin S$ joins S, the increase in the cost is $m_i(S) = c(S \cup \{i\}) - c(S)$. Taking account of the order in which the agents other than i join the full coalition N, there are $s!\,(n-s-1)!$ of $n!$ cases where agent i joins S. Then, the Shapley value, x_i^S, is defined as the weighted average of $m_i(S)$ over all coalitions $S \subseteq N \backslash \{i\}$, that is,

$$x_i^S = \sum_{S \subseteq N \setminus \{i\}} \frac{s!\,(n-s-1)!}{n!} m_i(S). \tag{9.1}$$

The A–S rule may be considered a generalization of the Shapley value to a continuous version (Hougaard 2009, p. 100). Suppose that there are n heterogeneous goods. Agent $i \in N$ demands good i, the quantity of which is q_i. Let $C(Q)$ denote the associated cost with the demand profile $Q = (q_1, \ldots, q_n)$. Then, the cost share of agent i, x_i^{AS}, is defined by the A–S rule as

$$x_i^{AS} = \int_0^{q_i} \frac{\partial C}{\partial q_i}\left(\frac{t}{q_i} Q\right) dt = q_i \int_0^1 \frac{\partial C}{\partial q_i}(tQ)dt. \tag{9.2}$$

Note that $\sum_{i=1}^n x_i^{AS} = C(Q)$.

If the cost function $C(Q)$ were at hand, the computation in (9.2) would be straightforward. However, in reality, some elaboration is necessary for the computation because any explicit expression of $C(Q)$ is unlikely to be available.

As an application of the A–S rule, let us consider a seminal work by Billera et al. (1978) on internal telephone billing rates. A monthly collection of phone calls made at a university is broken down into types that form the basis of the monthly billings, such as the time of day, destination, and type of day (weekday or weekend). Then, a price per minute of calling is to be set for each type, but this is not a straightforward task because the marginal cost of calling generally varies as total minutes of calls accumulate during the month. In this case, let $C(Q)$ denote the minimal cost of the monthly calls $Q = (q_1, \ldots, q_n)$, where q_i represents the total monthly minutes of calls of type i made by different departments of the university. This pricing rule embodies not only efficiency in that the total billings cover the total costs but also fairness in that a unique price is applied to all the calls of the same type.

9.2.2 Applications to the Relevant Problem

Several studies have applied the A–S rule to problems related to our present problem. These include Lin et al. (2005) and Lin et al. (2006b). They supposed that real and reactive power are supplied by generators. An independent system operator (ISO) must procure an adequate amount of reactive power support from generators; it compensates generators by making it indifferent for them to generate real power or provide reactive power.

Those studies defined the cost of reactive power as opportunity costs, that is, the profit a generator would have made if it had supplied real power instead of reactive power. Specifically, both studies above posited a cost function of real power production to form a reactive power cost as follows. On the one hand, Lin et al. (2005) assumed that a generator bids a marginal cost for real power supply. On the other hand, Lin et al. (2006b) assumed that a generator adds a certain rate of profit to the production costs. The ISO calculates the least costs of reactive power supply subject to such constraints as real and reactive power flow equations, power output limits, and voltage limits. Then, the authors applied the A–S rule to the allocation of the cost among consumers.

9.3 Calculation Methods

The present study proposes to allocate ancillary service costs to distributed RES-E generators because, as explained in Sect. 9.1, they owe the RES-E being fed into the grid to the ancillary services a system operator provides. This section focuses on a procedure for applying the A–S rule to our problem of how to allocate the ancillary service cost among distributed RES-E generators. However, the procedure for calculating the total ancillary service costs is left to other studies, such as Lin et al. (2005) and Lin et al. (2006b).

9.3.1 A Method of Repeated Optimization

The method presented in this section has been used so far by, for example, Billera et al. (1978) and Lin et al. (2005). Suppose that a distributed generator $i (i = 1, \ldots, n)$ could feed an amount of electricity q_i into the grid during a period of time. A certain set of electricity $Q = (q_1, \ldots, q_n)$ was successfully supplied as a whole during that period, owing to the ancillary service provided by a system operator. The cost of the ancillary service is defined as a function of Q, $C(Q)$. The system operator must allocate $C(Q)$ among the distributed generators. The A–S rule yields the cost share of generator i, x_i (Q), as presented in Eq. (9.2).

Next, calculate the integral in (9.2). Referring to Billera et al. (1978), we can compute it numerically as an approximate value in four steps. First, set $t_0 = 0$ and increase t from t_0 to t_1 by a fixed amount $\Delta t = 1/m$, where m is a somewhat large number. Compute $C(t_1 Q)$. Second, setting $t_2 = t_1 + \Delta t$, compute $C(t_2 Q)$. Repeat this procedure until t reaches 1. Third, compute the numerical partial derivatives, which are then integrated from 0 to 1. Finally, multiplying the resultant value by q_i yields generator i's share of the ancillary service cost, as Eq. (9.2) indicates.

9.3.2 A Method of Data Envelopment

The method briefly explained in Sect. 9.3.1 involves quite a few computations. An alternative method uses parametric linear programming, which is presented in this section.

A data envelopment technique may be applied to the computation of ancillary service cost allocation in the A–S rule (Hougaard and Tind 2009; Hougaard 2009). Referring to Hougaard (2009), we explain this method. For a given Q, set $C = C(Q)$. Then, (Q, C) constitutes the possibilities of an electricity output vector and its associated ancillary service costs. Let Z denote the set of these possibilities. It is postulated that Z satisfies the three conditions of convexity, decreasing returns, and free disposability: convexity means that if $z, z' \in Z$, then $\lambda z + (1 - \lambda)z' \in Z$ for $\lambda \in [0, 1]$; decreasing returns mean that if $z \in Z$, then $\lambda z \in Z$ for $\lambda \in [0, 1]$; and free disposability means that if $Q' \leq Q$ and $C' \geq C$ for $z = (Q, C) \in Z$, then $z' = (Q', C') \in Z$. Given a set of data $\Omega = \{(Q_i, C_i)\,|i = 1, \dots k\}$, we may construct an empirical estimate Z^* of Z, which can be used to calculate (9.2).

This technique may be directly applied to our problem as follows. Assume that the demand for electricity does not vary greatly. The problem is to allocate the ancillary cost, C_i, in period $i = 1, \dots, k$ between the n distributed generators. Suppose that the RES-E output profile in period i is expressed as $Q_i = (q_{i,1}, \dots, q_{i,n})$. Consider (tQ_i, tC_i), where $t \in [0, 1]$. For any t, calculate $\lambda_1, \dots, \lambda_k$ so that $\sum_{i=1}^{k} \lambda_i C_i$ is minimized subject to $\sum_{i=1}^{k} \lambda_i Q_i \geq tQ_i$. Then, by parametric linear programming, we may identify a series of intervals with respect to t; in each interval, the objective function is expressed as a linear function of $tq_{i,1}, \dots, tq_{i,n}$, which we can use to calculate the cost share of a distributed generator from (9.2).

Recall that the technique explained in Sect. 9.3.1 calculates $C(tQ)$ by calculating the ancillary service costs repeatedly. On the other hand, the technique in this section constructs $C(tQ)$ from observations of RES-E output and the associated ancillary service costs by parametric linear programming. This technique, if available, would be more tractable than the first technique in applying the A–S rule to the problem of allocating ancillary service costs.

The technique in this section is deeply dependent on the assumption that the three conditions, i.e., convexity, decreasing returns, and free disposability, are satisfied. If the demand varies greatly among the observations, the conditions might not hold and the technique might not be applicable. In this case, it might be necessary to use the first technique presented in Sect. 9.3.1.

9.4 Conclusion

This chapter presented a method of applying the A–S rule to the allocation of ancillary service costs to grid-connected RES-E generators. In particular, the use of a data envelopment technique was proposed to calculate each generator's cost share.

It is more likely to be necessary to allocate ancillary service costs to distributed RES-E generators. Technological developments and various policy support measures have enabled an increasing amount of RES-E generated by distributed generators to be fed into the power grid. Governments strongly promote RES-E generation because it has many advantages, such as increasing energy supply, reducing greenhouse gas emissions, and improving energy security. On the other hand, at the same time, the RES-E fed into the grid may cause various problems, such as frequency and voltage instability, in an electrical power system. Accordingly, appropriately sharing ancillary service costs will be increasingly important. The proposed method of applying the A–S rule offers a new way to resolve the problem of allocating ancillary service costs to distributed RES-E generators.

The method proposed for computing the cost allocation by means of the A–S rule is strongly dependent on the three conditions, i.e., convexity, decreasing returns, and free disposability. Thus, it is important to examine whether the ancillary service costs satisfy these conditions. If some of them satisfy the conditions and others do not, the ancillary services should be divided into those that satisfy the conditions and those that do not. While the data envelopment technique may be applied to the first type of ancillary service costs, the method of repeated optimization should be applied to the second type.

There are many households and businesses that undertake small-scale distributed generation, such as PV generation on the rooftops of houses and stores. One possibility is for these small-scale generators not to participate in the proposed scheme but rather to incur a fixed cost per unit of RES-E they feed into the grid. Alternatively, they may be grouped together and participate in the proposed scheme as a single entity.

The concept in which individual RES-E generators share ancillary service costs may constitute a new type of feed-in tariff (FIT) system. A FIT system, as is well known, allows distributed generators to feed the RES-E they produce into the power grid and sell it at a favorable price. This policy instrument aims to promote RES-E generation to reduce greenhouse gas emissions, increase energy supply, improve energy security, and provide other benefits. Thus, the use of such an instrument might no longer be necessary after a certain goal of diffusion is achieved. On the other hand, those who undertake RES-E generation want to sell the RES-E they produce at a favorable price, as in the current FIT system. In this sense, too, a new type of FIT should succeed the current FIT. The present approach would contribute to designing this new type of FIT system.

To cope with the unstable nature of RES-E generation, typically seen in solar PV and wind-power generation, installations of power storage systems are gradually increasing. Because it is still very costly for a small-scale generator to install such a system on its own, it is more likely that such a system will be owned by a group of distributed RES-E generators. In this case, the problem of cost allocation or benefit distribution will once again arise in the group. The proposed method may be applied to this case as well.

Technological aspects are crucial for the proposed method to be practically adopted in an electrical power system. In particular, a system operator must compute and procure an optimal level of ancillary services. Information technology may play a role in this regard, and a smart grid system is one promising technology. Collaborative studies that include relevant research fields are of great importance going forward.

References

Aumann RJ, Shapley LS (1974) Values of non-atomic games. Princeton University Press, Princeton

Billera LJ, Heath DC, Raanan J (1978) Internal telephone billing rates: a novel application of non-atomic game theory. Oper Res 26(6):956–965

Choi JY, Rim S-H, Park J-K (1998) Optimal real time pricing of real and reactive powers. IEEE Power Syst 13(4):1226–1231

Hougaard JL (2009) An introduction to allocation rules. Springer, Berlin

Hougaard JL, Tind J (2009) Cost allocation and convex data envelopment. Eur J Oper Res 194:939–947

Kirschen D, Strbac G (2004) Fundamentals of power system economics. Wiley, West Sussex

Koide A, Tsuji T, Oyama T, Hashiguchi T, Goda T, Shinji T, Tsujita S (2012) A study on real-time pricing method of reactive power in voltage profile control method of future distribution network. IEEJ Trans Power Energ 132(4):359–370 (In Japanese with the English abstract)

Lin XJ, Yu CW, Chung CY (2005) Pricing of reactive support ancillary services. IEE Proc-C 152(5):616–622

Lin XJ, Yu CW, David AK, Chung CY, Wu H, Xu N (2006a) A novel market-based reactive power management scheme. Adv Mater Res-Switz 28:127–132

Lin XJ, Yu CW, Xu N, Chung CY, Wu H (2006b) Reactive power service cost allocation using Aumann-Shapley method. IEE Proc-C 153(5):540–546

Stoft S (2002) Power system economics. Wiley–IEEE Press, New York

Ueki Y, Hara R, Kita H, Hasegawa J (2008) Evaluation of contribution for voltage control ancillary services based on social surplus. IEEJ Trans Power Energ 128(10):1243–1251 (In Japanese with the English abstract)

Chapter 10
Opinion Leadership in the Diffusion of Photovoltaic Systems

Abstract According to diffusion theory, opinion leaders—through interpersonal communication with potential adopters—play an important role in the diffusion of new technologies. The purpose of this chapter is to examine whether this is the case for a solar photovoltaic (PV) power generation system and to investigate the role and utility of opinion leadership in the diffusion of PV systems. Specifically, the study proposes, examines, and considers the implications of the hypothesis that there is a positive relationship between willingness to pay (WTP) for a PV system and opinion leadership on the adoption of PV systems in the residential sector. The study employed an internet-based questionnaire survey to assess the use of interpersonal communication in decision-making on adoption, to identify opinion leaders with respect to adoption and to characterize their WTP. The response pool consisted of 488 individuals who lived in detached houses in Japan, owned a residential PV system and were responsible for making the decision to adopt their PV system. The results support the hypothesis stated above. Considering that subsidization preferentially incentivizes households with greater WTP to adopt PV systems, the results suggest that subsidization is more effective than purchases of PV electricity under a feed-in tariff system in promoting the diffusion of PV systems in the residential sector through interpersonal communication.

Keywords Residential photovoltaic system · Diffusion · Opinion leadership · Willingness to pay · Questionnaire survey

10.1 Introduction

Governments seeking to diffuse photovoltaic (PV) systems widely throughout society typically offer incentives such as subsidization, tax credits, and feed-in

---

Modified, with permission of Elesevier, from Yamamoto, Y., Opinion leadership and willingness to pay for residential photovoltaic systems, Energy Policy, 83, 185–192, Elsevier, 2015. I would like to thank Elsevier.

Y. Yamamoto, *Feed-in Tariffs and the Economics of Renewable Energy*,
https://doi.org/10.1007/978-3-319-76864-9_10

tariffs (FITs) to promote widespread adoption. Such incentives, along with the reductions in the cost of PV systems that have accompanied the development of PV technology, increase the advantages of adopting PV systems for potential adopters. Generally, if the advantages that a new technology offers include economic advantages over existing technology, the rate of adoption will be rapid (Rogers 2003, pp. 229–230). On the basis of this understanding, numerous studies have investigated the diffusion of PV systems, including Lesser and Su (2008), Parker (2008), and Rigter and Vidican (2010), among others. According to the view of technology diffusion reflected in these studies, the key to successful diffusion is cost reduction through technological development and through incentives provided by the government. Many studies have further investigated the diffusion of PV systems with regard to FITs; recent examples include Cherrington et al. (2013), Martin and Rice (2013), Antonelli and Desideri (2014), and Campoccia et al. (2014).

However, another perspective on technology diffusion—one with implications for the diffusion of PV systems in particular—also merits attention. This perspective focuses not on the new technology to be diffused but rather on the members of society who adopt it. Studies based on this understanding include Faiers and Neame (2006), Bollinger and Gillingham (2012), Zhai and Williams (2012), Rai and Robinson (2013), Noll et al. (2014), and Schelly (2014). Implicit in this view is the epidemic model, in which it is presumed that "the primary factor limiting diffusion is information, and that the most important source of information about a new technology is people or firms who have tried it" (Jaffe et al. 2003, p. 489). From this viewpoint, communication between those who have already adopted and those who have not yet but are considering adopting must be understood as playing an important role in diffusion. Furthermore, such communication is most likely to occur if those who have already adopted are regarded as opinion leaders who are often asked for information or advice.

In a seminal work on technology diffusion, Rogers (2003) offered a comprehensive analysis of the role played by opinion leadership and opinion leaders in the diffusion of innovations such as new ideas and technologies. According to Rogers (2003), diffusion may be defined as the process in which an innovation is communicated among the members of a social system (p. 5). Opinion leadership is then the degree to which an individual is able to influence other individuals' attitudes or behavior, and an opinion leader is such an individual (p. 300). From these definitions, it follows that information or advice from an opinion leader is helpful for potential adopters to decide the adoption of an innovation.

This view suggests that if opinion leaders adopt PV systems during an early phase of the diffusion of a new technology, the subsequent diffusion—through the mechanism of interpersonal communication—will be broader, more rapid, or both than it would be if they had not adopted. The assumption of interpersonal communication as the mechanism of opinion leadership furthermore suggests an alternative policy approach to incentives than is currently employed to promote the diffusion of PV systems. If some characteristic identifying a person as an opinion leader on adoption could be identified, a government might use it in designing an incentive program to encourage opinion leaders to preferentially adopt. Critically, the development of such an approach would depend on finding such a characteristic.

We hypothesize that there is a positive relationship between willingness to pay (WTP), that is, the maximum amount of money a person is willing to pay for a PV system, and the level of opinion leadership exerted by that person, and hence, that greater WTP can serve as a condition indicating higher opinion leadership. Note that WTP can be understood to represent how much a person values a PV system because he or she will not pay more than this value to obtain one. The reasoning behind this hypothesis is that opinion leaders are more likely than others to recognize—and pay for—the high value that solar power generation by PV systems offers to society.

If this hypothesis is correct, then policymakers could increase the diffusion of PV systems with programs designed to incentivize those with greater WTP to preferentially adopt. This would be a relatively easy task because WTP involves monetary values that could tractably be reflected in an incentive program.

This chapter offers an empirical investigation that supports the above hypothesis. In the investigation, the first question examined was whether interpersonal communication plays a role in decision-making about the adoption of a PV system. In a survey of households that already had PV systems, respondents were asked, first, whether they had received information or advice from those who had already adopted, and second, whether they were asked for information or advice from those who were considering adopting. When it was shown that interpersonal communication plays a role in decision-making about adoption, the investigation then turned to characterizing respondents with regard to opinion leadership and WTP. In particular, opinion leaders were identified as those who had not only adopted the innovation—in this case, a PV system—earlier than others but also were respected enough by other members of a social system to be asked for information or advice about the innovation.

The remainder of this chapter is organized as follows. Section 10.2 presents, as background, a brief review of preceding studies, as well as a brief survey of the status of PV system diffusion and governmental policy with regard to diffusion in Japan. Section 10.2 also presents the methodology of the survey and an overview of the questionnaire; critically, this part explains how opinion leaders are identified. Section 10.3 presents the results and offers a discussion regarding the use of interpersonal communication in the diffusion of PV systems, identification of opinion leaders, and WTP. Finally, Sect. 10.4 explores the policy implications of the results and concludes with suggestions for future research.

10.2 Methods

As background to the present study, which employs a questionnaire survey conducted in Japan, this section begins with a brief review of several previous questionnaire-based studies, as well as a survey of the status of PV system diffusion and governmental policy with regard to diffusion in Japan. (For more on the diffusion of PV systems in Japan, see also Mendonça 2007, pp. 71–75.)

Critical to the survey employed in this study is the methodology of how opinion leaders are identified among respondents. The latter half of this section describes the procedure employed for identifying opinion leaders, which is predominantly based on the work of Rogers (2003). Finally, this section offers an overview of the survey, including a description of the respondents.

10.2.1 Literature Review

Numerous studies have investigated the role of interpersonal communication in technology diffusion, among which Griliches (1957) and David (1966) are widely regarded as seminal. Other studies have addressed energy issues (Stern 1992; Wilson and Dowlatabadi 2007; Bollinger and Gillingham 2012; Rai and Robinson 2013). However, only a limited number of studies have focused on the diffusion of residential PV systems. Two of these in particular support the present study by showing the role of interpersonal communication in PV technology diffusion: Bollinger and Gillingham (2012), which provided a methodology for identifying peer effects and applied it to the diffusion of PV systems in California, and Rai and Robinson (2013), which quantified the effects of different information channels on aspiring PV adopters' decision-making.

Furthermore, the following studies provide context for the questionnaire survey-based approach taken in the present study. First, Labay and Kinnear (1981) investigated, by means of a questionnaire survey, differences in perceptions of the attributes of residential solar heating and hot water systems between adopters and non-adopters, with attention to the demographic characteristics of respondents. Like the present study, Labay and Kinnear examined decision-making in the purchasing of a new technology, though not PV systems, in light of differences among people. In contrast, the present study is concerned not with differences between adopters and non-adopters—Labay and Kinnear's concern—but rather with differences in levels of opinion leadership among adopters.

Second, Jager (2006) also employed a questionnaire survey to investigate how people reach the decision to adopt a PV system. Jager examined the effects of information and support meetings organized by a municipality. That study shed light on the social aspects of diffusion, showing that social motives, such as belongingness and participation, were more important to those who invested less cognitive effort in the initial phase of the decision-making process. Similarly, Noll et al. (2014) investigated the roles of community organization and peer effects in the adoption of residential PV systems. The present study, in contrast, is concerned with a different social aspect of diffusion, namely, interpersonal communication between adopters and potential adopters of PV systems.

Third, Shirai et al. (2012) employed a questionnaire survey to compare the characteristics of adopters and non-adopters of PV systems in Japan. In their study, adopters were segmented on the basis of the year they adopted, while non-adopters were divided between those who were considering adoption (potential adopters) and those who were not (unlikely adopters). Shirai et al. showed that the earlier the year of adoption, the higher adopters scored compared to unlikely adopters in terms of awareness and leadership on environmental problems, innovativeness, community involvement, and participation in volunteer work or club activities. Like Labay and Kinnear (1981), Shirai et al. compared actual or potential adopters and unlikely adopters, in contrast to the present study, which examines differences among adopters. One of the findings of Shirai et al., however, has great significance for the present study: the researchers found that those who adopted earlier were more likely to be concerned about environmental issues and to exercise opinion leadership on those issues. As will be explained in Sect. 10.2.3, the present study identifies opinion leaders on the adoption of a PV system in part based on the year in which they adopted.

Finally, Schelly (2014) interviewed individuals who had adopted residential PV systems to explore what had motivated them to adopt. Based on the findings of the study, Schelly argues that the adoption of technological innovations is promoted to some extent through informal information sharing, that is, the desires of individuals to educate others by demonstrating the feasibility of a technology or setting an example for a community. Schelly's notion of informal information sharing corresponds, to a certain degree, with opinion leadership in the present study.

10.2.2 Diffusion of PV Systems and Policy in Japan

As shown in Fig. 10.1, the number of households that have adopted PV systems in Japan has increased annually, with the exception of the three-year period from 2006 through 2008 (where years cited, here and henceforth, refer to the Japanese fiscal year). This upward trend is largely due to reductions in the cost of adoption, which are partly attributable to ongoing progress in the development of PV technology. Figure 10.2 shows the gradual decrease in the cost of PV systems, including PV panels, other equipment, and installation, in Japan.

Government-provided incentives, including investment subsidies and feed-in tariffs, have also contributed to reductions in the cost to households of PV systems. Households in Japan typically receive investment subsidies from the national government for purchasing and installing PV systems. For example, in 2013, the Japanese government provided a subsidy of ¥15,000 or ¥20,000, depending on the cost of purchasing and installing the system, per kW of installed capacity of PV panels. (¥100.00 was approximately equivalent to $1.00 at the time.) The subsidization of residential PV system adoption was first introduced at the national level in Japan in 1994. Notably, subsidization was suspended in March 2006, then reinstated in January 2009, a period roughly corresponding to the interruption—from 2006 through 2008—of the upward trend in adoption, as shown in Fig. 10.1.

In addition to subsidization by the national government, many prefectures and municipalities in Japan offer investment subsidies. For example, the Gunma prefectural government offered a subsidy of ¥15,000, and the city of Takasaki, Gunma,

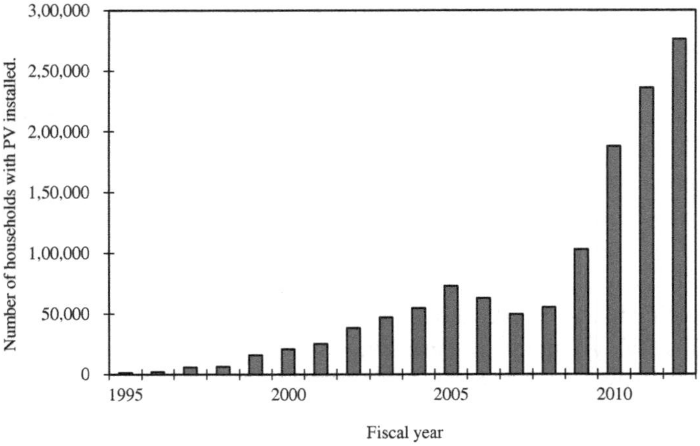

Fig. 10.1 Annual adoption of residential PV systems. *Data* JPEA (2014)

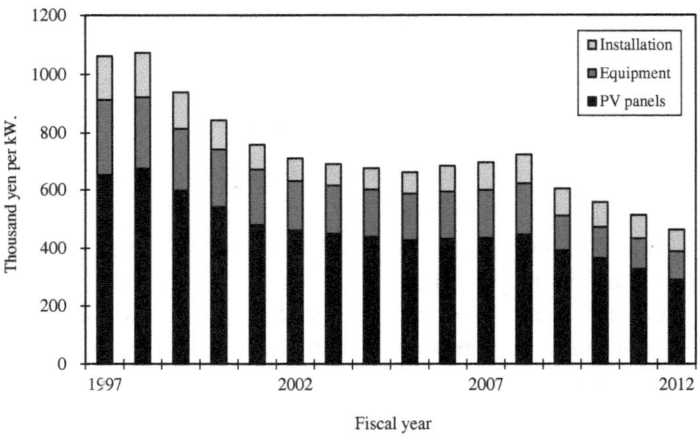

Fig. 10.2 PV system costs (panels, equipment, and installation). *Data* METI (2013)

a subsidy of ¥30,000, per kW of installed capacity in 2013. Thus, three levels of government offered subsidies to households in Takasaki, Gunma, in 2013.

Moreover, Japan also employs a FIT system, under which households sell surplus PV electricity, i.e., electricity generated by residential PV systems but not consumed by the household, to the electric utility at a set price for either 10 or 20 years, depending on the installed capacity of the PV system. In 2013, the PV electricity price paid by the electric utility was ¥38.0 per kWh for PV systems smaller than 10 kW installed capacity, and it was ¥37.8 per kWh for larger systems. These prices were significantly higher even than retail electricity rates, which were ¥18.89 per kWh for use up to 120 kWh, ¥25.19 from 120 to 300 kWh, and ¥29.10 for use above 300 kWh (Tokyo Electric Power Company, June 2013).

Historically, electric utilities began to voluntarily purchase PV electricity in Japan in 1992, paying for surplus PV electricity at a rate nearly equal to the retail electricity rate at the time. In November 2009, electric utilities were required by law to purchase PV electricity at ¥48.0 per KWh for 10 years. Recall that 2009 was the year in which national subsidization resumed and that the rate of adoption also resumed its rapid growth. Japan's current FIT system was implemented in July 2012, replacing a renewable portfolio standards (RPS) system that had been in effect since April 2003; notably, households were able to sell surplus PV electricity to utilities even under the RPS system.

10.2.3 Procedures for Identifying Opinion Leaders

A sociometric approach, in which people are directly asked to identify opinion leaders in a community, may be an effective way to identify opinion leaders (Rogers 2003, pp. 308–310; Kramer et al. 2009). However, this approach was not practical for the present study because it would limit the number of subjects to those who could be interviewed. Accordingly, this study used a questionnaire survey, as described in Sect. 10.2.4, as an alternative way to identify opinion leaders. This section explains the procedure employed to identify opinion leaders through the questionnaire survey.

Generally speaking, opinion leadership can be categorized in terms of polymorphism and monomorphism, depending on the number of topics on which an opinion leader has influence. Polymorphism is concerned with a variety of topics, whereas monomorphism is concerned with a single topic (Rogers 2003). The present study is concerned with monomorphic opinion leadership: leadership specifically on the topic of the adoption of a PV system. Hence, for our purposes, it is sufficient to identify opinion leaders only with regard to the adoption of a PV system.

Two criteria were employed in this study to identify opinion leaders on the adoption of PV systems. The first was the time of adoption: earlier adopters have a higher degree of opinion leadership than other adopters (Rogers 2003, p. 291). Shirai et al. (2012), as we discussed in Sect. 10.2.1, offered support for this view: in measuring opinion leadership on environmental issues, the researchers showed that those who had adopted PV systems earlier tended towards higher opinion leadership. Key to the present study, therefore, was to determine the year before which those who adopted could be considered earlier adopters. The analytical method employed for making this determination is discussed in Sect. 10.3.2.

This first criterion, however, does not suffice on its own for identifying opinion leaders on adoption because not all who adopted earlier are necessarily opinion leaders. As Rogers (2003) proposes, five adopter categories can be distinguished based on the time of adoption. The five categories, from earlier to later, are innovators, early adopters, early majority, late majority, and laggards. Among these, an individual in the category of early adopter has the highest degree of opinion leadership. In contrast, an individual in the category of innovator, despite being an earlier adopter, may not be respected by other individuals. Accordingly, a criterion in addition to time of adoption is needed.

As the second criterion, this study employed self-identification, or "self-designation" as Rogers (2003) terms it, to identify opinion leaders on the adoption of PV systems. Assessment of self-identification as an opinion leader, as Rogers argues, is appropriate when interrogating respondents and is a task for which a questionnaire survey is better suited than other investigatory techniques such as sociometric methods, informants' ratings, and observation. To assess self-identification, respondents are asked to indicate the degree to which others regard them as influential by answering a question such as the following, "Do you think people come to you for information or advice more often than to others?" (pp. 308–312). To assess self-identification with regard to opinion leadership on the adoption of a PV system, the questionnaire survey in this study included the question, "After installation, were you asked for information or advice about PV systems by anyone who was considering adoption?"

It should be noted that there are different views on adopter categories and their relationships. Moor (2014) argued that a chasm separates the early adopters and the early majority; however, this is not relevant to the present study, which deals only with earlier adopters. On the other hand, Sultan and Winer (1993) argued that time preferences for a product differ depending on the attributes of a product: an innovator for one product may be a laggard for another; this point is also irrelevant to our study, which is concerned with monomorphic opinion leadership.

After opinion leaders on the adoption of PV systems were identified by the above two criteria, identified leaders were tested for characteristics that opinion leaders in general are often thought to have to a high degree. According to Rogers (2003), these include external communication, accessibility, socioeconomic status, innovativeness, and responsiveness to social norms (pp. 316–319). Three characteristics in particular—accessibility, socioeconomic status, and innovativeness—are relevant with regard to the adoption of PV systems in Japan. First, accessibility is related to how socially accessible opinion leaders are; a possible indicator of this characteristic is the frequency of participation in social activities such as community, volunteer, or club activities. Second, socioeconomic status refers to income and social recognition. Third, innovativeness refers to the degree to which an individual adopts innovations earlier than other members of a social system. Section 10.3.2 discusses the analytical method for assessing the first two of the three characteristics in those identified as opinion leaders based on time of adoption and self-identification. Innovativeness, the third characteristic, does not need to be assessed empirically because it is, by definition, a characteristic of earlier adopters, and earlier adoption is the first criterion by which opinion leaders on the adoption of PV systems were identified.

The remaining two characteristics—external communication and responsiveness to social norms—are less relevant to the situation in Japan, where the survey was conducted. External communication pertains to bringing an idea from outside into the social group to which an individual belongs. In Japan, however, people are typically introduced to the idea of PV power generation as an innovation not through communication with individuals but rather, predominantly, through the mass media. The characteristic of responsiveness to social norms could be significant in that opinion leaders decide whether to adopt an innovation as part of their conformity to social norms. Rogers (2003, p. 318) finds that opinion leaders are more innovative if a social system's norms favor change; otherwise, they are not. In Japan, however, PV power generation is widely accepted as an innovation that contributes to society, such that the norms of the social system favor its introduction.

10.2.4 Overview of the Questionnaire Survey

For the present study, an internet-based questionnaire was employed in a survey of randomly selected subjects. A research company conducted the survey in Japan using its own subject pool, which it maintains as follows. First, internet users seeking to join the pool register their name, gender, date of birth, address, phone number, income level, occupation, and other characteristics on the company's website, thereby making themselves available for the various surveys the company conducts on behalf of researchers. Second, the company invites randomly selected members to take a survey; the invited members are paid if they respond to the survey. To maximize the reliability of the subject pool, whenever it conducts a survey, the company checks the responses that each invited member provides against his or her registration information in order to detect inconsistencies. If a member is identified as unreliable, he or she is removed from the pool. The company has been maintaining the pool in this manner for more than ten years and over 6000 surveys, so the subject pool may be regarded as highly reliable.

When the survey was conducted, the pool consisted of 728,081 subjects living in Japan. Although the profile of the pool was different from that of the Japanese people as a whole (Fig. 10.3), this internet-based survey enabled random selection of subjects who owned residential PV systems at home; such a selection would have been impossible if a conventional survey method had been used.

The survey comprised two steps. The first step was a screening process that eliminated all but those respondents who lived in a detached house, owned a residential PV system and were responsible for making the decision to adopt a PV system. For this step, 334,630 subjects were randomly selected from the pool, and 63,580 responses (19.0%) were obtained (Table 10.1). The number of those who satisfied the above three conditions was 2810 of 63,580 respondents (4.4%). The percentage 4.4% is comparable to 4.7% of all detached houses having residential PV systems in or before 2012 in Japan (2013, Statics Japan). After 182 invalid responses were removed, 2628 respondents remained.

The second step was the completion of the questionnaire by the 2628 respondents obtained in the first step. The questionnaire survey was conducted in April 2013. It yielded a total of 558 responses (21.2%), twelve of which contained obvious errors, probably due to mistyping. Thus, 546 responses (20.8%) were obtained in the second step. The 546 respondents were, as a group, similar in profile—with respect to gender, age, household income, and education—to those who did not respond or did respond but had invalid responses (Table 10.2).

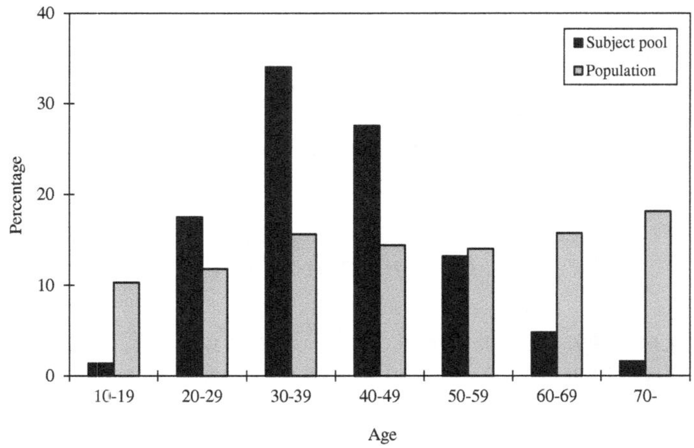

Fig. 10.3 Comparison between the subject pool and the Japanese population with respect to age distribution. *Data* Statistics Bureau of Japan (2010)

Table 10.1 Participation in process of obtaining the response pool

	Size	Percentage		
Subject pool	728,081	100.0		
Invited to 1st survey	334,630	46.0	100.0	
Completed 1st survey	63,580		19.0	
Qualified for and sent 2nd survey	2628		0.8	100.0
Completed 2nd survey	558			21.2
Invalidated	(12)			(0.5)
Obtained from 2nd survey	546			20.8
Excluded as explained in 10.2.4	(58)			(2.2)
Response pool	488			18.6

Initial review of the 546 responses indicated the necessity of three further restrictions of the response pool. First, there were 19 respondents who did not respond to the question of how much they spent on a PV system because they had adopted their system when purchasing or remodeling a house, and the cost of the system was not broken out of the total bill; these respondents were excluded from the response pool. Second, there were a number of respondents who had paid unusually small or large amounts of money for their PV systems compared to the typical amount spent on

Table 10.2 Comparison of respondents and non-respondents

	Percentage	
	Respondents ($n = 546$)	Non-respondents ($n = 2082$)
Gender		
Male	71.2	70.8
Female	28.8	29.2
Age		
20–29	4.2	4.0
30–39	29.3	26.9
40–49	29.3	32.2
50–59	19.6	21.8
60–69	13.2	11.5
70–	4.4	3.5
Household income (Million yen)		
–2.00	3.1	3.5
2.01–4.00	13.4	14.5
4.01–6.00	28.4	26.4
6.01–8.00	23.4	25.1
8.01–10.00	15.8	15.9
10.01–	15.9	14.8
Education		
High school	35.2	37.7
University or college	55.1	54.2
Graduate school	9.7	8.1

Non-respondents include 12 whose responses were invalidated

such a system; respondents who had paid fewer than ¥0.10 million (6 respondents) or more than ¥10.00 million (1 respondent) were also excluded from the response pool. These cut-off values were determined based on the average amount of spending (Fig. 10.2) and an average capacity of 4 kW for a residential PV system in Japan. After these exclusions, a total of 520 responses remained. Third, there were also respondents whose PV systems had unusually smaller or larger capacities than residential systems generally have in Japan. Respondents whose system capacity was smaller than 2.0 kW (6 respondents) or larger than 6.0 kW (26 respondents) were both excluded from the response pool. These cut-off values were set based on the average ($\mu = 4.00$) and standard deviation ($\sigma = 1.37$) of the PV capacity over the 520 responses. That is, they were set at $\mu \pm 1.5\sigma$, which corresponds to 2.0 kW and 6.0 kW in the data set, covering 93.4% of the 520 responses. After these three restrictions, a total of 488 responses remained.

Finally, the WTP reported on 164 of these responses was smaller than the amount the respondents had actually spent on their PV systems. This is an inconsistency if the assumption holds that people make rational purchase decisions. Hence, if a

Table 10.3 Responses on use of interpersonal communication

	Yes	No
Did you consult anyone?[a]	157 (32.2%)	331 (67.8%)
Did anyone consult with you?[b]	215 (44.1%)	273 (55.9%)

[a]Did you consult anycne who already owned a PV system before adopting?
[b]Did anyone thinking about adopting a PV system ask you for information or advice after you adopted?

respondent's reported WTP was smaller than the actual payment he or she made, then—based on an assumption of rational choice—the actual payment was substituted for the reported WTP in the analysis.

10.3 Results and Discussion

The survey results are analyzed to assess the role of interpersonal communication in decision-making about the adoption of a PV system, to identify opinion leaders among respondents, to determine opinion leaders' WTP and to examine the relationship between opinion leadership and WTP. The findings offer strong support for the hypothesis that there is a positive relationship between WTP for a PV system and level of opinion leadership on adoption.

10.3.1 Use of Interpersonal Communication

Analysis of the survey results shows that interpersonal communication plays a role in decision-making about the adoption of a PV system for a significant number of adopters. The number of respondents who, before adopting, consulted someone who already owned a PV system is 157 (32.2%), and the number of those who did not consult with anyone is 331 (67.8%), as shown in Table 10.3. In other words, one out of 3.1 respondents used interpersonal communication in making their decision to adopt.

In contrast, the number of respondents who, after adoption, were asked for information or advice by someone about a PV system is 215 (44.1%), and the number of those who were not is 273 (55.9%), as shown in Table 10.4. Thus, one out of 2.3 respondents received an inquiry about PV systems after they had adopted.

Notice that the percentage of those whose advice was sought is higher than the percentage of those who sought advice. This means that a relatively large number of respondents played a role in disseminating information or advice about PV systems, in turn suggesting that these respondents had a relatively high level of opinion leadership.

Table 10.4 Comparative results for PV opinion leaders and other respondents

	Percentage	
	PV opinion leaders ($n = 73$)	Other respondents ($n = 415$)
Participates in social activities	71.2	49.2
Annual income higher than ¥8.00 million	42.5	30.1
Generally asked for information or advice by others often	46.6	25.1
Offers information on PV to others via the internet	21.9	13.5
Values environmental benefits of PV more than economic advantages	30.1	20.7

10.3.2 *Identification of Opinion Leaders*

As discussed in Sect. 10.2.3, respondents were assessed for opinion leadership. First, they were assessed on the basis of two criteria with regard to PV system adoption: time of adoption and self-identification; next, respondents meeting these criteria (henceforth designated "PV opinion leaders") were tested for characteristics common among opinion leaders in general.

With regard to time of adoption, the first analytical task is to set a cut-off year beyond which a respondent would not be considered as an earlier adopter. As mentioned in Sect. 10.2.2, national subsidization of PV system purchases was resumed, and electric utilities were required by law to purchase PV electricity, in 2009. This may have contributed to the upward trend in adoption beginning that year (Fig. 10.1). Hence, it is reasonable to consider households that adopted no later than 2008 as earlier adopters. Accordingly, the year 2008 was set as the cut-off year. A total of 161 respondents (33.0%) adopted no later than 2008 and thus qualify as earlier adopters.

The second criterion for identifying PV opinion leaders, self-identification, was assessed by asking respondents whether they were asked for information or advice about a PV system by others after they had adopted. As discussed in Sect. 10.3.1, a total of 215 respondents (44.1%) answered yes to this question. A total of 73 respondents (15.0%) met both of the two criteria and were thus identified as PV opinion leaders.

Next, PV opinion leaders were assessed for several general characteristics of opinion leaders discussed in Sect. 10.2.3, namely, accessibility and socioeconomic status (Table 10.4). With regard to accessibility, which can be measured by social participation, the respondents were asked whether they participated in social activities other than their jobs, such as community, volunteer, or club activities: 71.2% of PV opinion leaders participated in such activities, compared with 49.2% of other respondents (z test for two proportions: $z = 3.48$, $p < 0.01$). Thus, PV opinion leaders may be regarded as demonstrating significantly greater accessibility than other respondents.

Socioeconomic status was measured by annual income level. The results showed that the annual income of the PV opinion leaders identified was significantly higher than that of other respondents: 42.5% of PV opinion leaders reported their annual income as higher than ¥8.00 million, while only 30.1% of other respondents reported incomes at that level ($z = 2.09$, $p < 0.05$).

As discussed in Sect. 10.2.3, an additional relevant characteristic of opinion leaders, innovativeness, is necessarily a quality of all PV opinion leaders identified because earlier adopters are by definition innovative, and all PV opinion leaders, by the first criterion discussed above, adopted their PV systems earlier than other respondents.

Hence, based on the characteristics common to opinion leaders in general, PV opinion leaders qualify more generally as opinion leaders.

Three other findings regarding PV opinion leaders are noteworthy (Table 10.4). First, 46.6% of PV opinion leaders were, in general, often asked for information or advice by someone, compared with 25.1% of other respondents ($z = 3.76$, $p < 0.01$). Second, 21.9% of PV opinion leaders shared information on PV systems via the internet, compared with 13.5% of other respondents ($z = 1.87$, $p < 0.05$). Lastly, 30.1% of PV opinion leaders believed that the benefits of PV power generation for the environment were more important than its economic advantages, compared with 20.7% of other respondents ($z = 1.79$, $p < 0.05$). The last finding in particular may reveal a possible reason that PV opinion leaders adopt PV systems in spite of high costs (Sect. 10.3.3 below), suggesting that they are deeply concerned about environmental issues.

10.3.3 Opinion Leaders' Willingness to Pay

The average WTP of PV opinion leaders was ¥2.54 million, whereas that of other respondents was ¥2.32 million (t test for two means: $t = 1.72$, $p < 0.05$). For PV opinion leaders, average WTP was thus greater—to a statistically significant degree—than that of other respondents.

This finding may be explained as follows. Recall that the first criterion employed to identify PV opinion leaders was the time of adoption because the earlier a respondent adopted, the more likely he or she was to be an opinion leader. Also note that according to a report published by the Ministry of Economy, Trade and Industry (2013), the cost that a household incurs in buying and installing a PV system has dropped consistently since 2009 (Fig. 10.2). Hence, the PV opinion leaders identified adopted when costs were still high. This may have led to PV opinion leaders' greater WTP.

An alternative, general explanation of this finding is also possible. As suggested in Sect. 10.3.2, and as Shirai et al. (2012) note as well, opinion leaders demonstrate a high level of concern about environmental issues, and they understand the importance of power generation from renewable energy sources in relation to problems such as global warming, energy security, and natural resource scarcity. This could explain the high value they place on solar power generation, represented by their greater WTP than others for a PV system. It should also be noted that opinion leaders tend to have a higher level of income than others, as mentioned in Sect. 10.2.3 and confirmed in Sect. 10.3.2; because people with higher incomes can afford to pay more, their WTP will be greater than that of people with lower incomes (Zhang and Wu 2012; Zorić and Hrovatin 2012; Guo et al. 2014).

10.3.4 Opinion Leadership in Relation to Willingness to Pay

In Sect. 10.3.3, the WTP of PV opinion leaders was compared to that of other respondents, and it was suggested that a higher level of opinion leadership implies greater WTP for a PV system. This subsection examines the converse, that is, whether greater WTP implies a higher level of opinion leadership. To answer this question, it is necessary, as a first step, to determine what level of WTP should be considered high WTP. As a second step, the proportion of PV opinion leaders among those with high WTP must be determined.

To complete the first step, let us determine a lower limit for high WTP. Respondents whose WTP is no less than this value may be considered, as a group, to have greater WTP than others. It is appropriate to set this lower limit such that approximately the top 25% of respondents by WTP are considered to have high WTP in view of 15.0% of respondents being PV opinion leaders. The proposed top 25% of residents by WTP corresponds to respondents who indicated WTP of no less than ¥3.00 million. The ratio of respondents with WTP of no less than ¥3.00 million to all respondents is 25.4% (124 of 488 respondents), which is acceptably close to the proposed lower limit by percentage. Hence, a lower limit of ¥3.00 million was set as the cut-off value for high WTP.

Let us proceed to the second step. The ratio of PV opinion leaders among respondents with WTP of at least ¥3.00 million is 25.0% (31 of 124 respondents); in comparison, the ratio of PV opinion leaders among respondents with WTP of less than ¥3.00 million is 11.5%, 42 of 364 respondents ($z = 3.63$, $p < 0.01$). Thus, the results show that greater WTP implies higher opinion leadership to a statistically significant degree.

This finding may be explained as follows. Note the difference in the arguments here and in Sect. 10.3.3 above. People with greater WTP for a PV system can afford to adopt even when the costs are high. Taking into account that the earlier the adoption, the higher the costs (Ministry of Economy, Trade and Industry 2013), it makes sense that people with greater WTP would adopt earlier. Recall that earlier adopters are, by definition, more likely to be PV opinion leaders.

The following may provide an alternative, general explanation. People with greater WTP for a PV system will be those with higher incomes or higher valuations of PV systems. In turn, highly valuing the contribution of PV power generation to solving environmental and energy problems is considered typical of opinion leaders. It should be noted that, in contrast to the discussion in Sect. 10.3.3, people with greater WTP may have higher incomes, but those with higher incomes are not necessarily opinion leaders.

10.4 Conclusion

This chapter shed light on the role and utility of interpersonal communication in the diffusion of PV systems in society. Based on the results of a questionnaire survey, the study demonstrated that people use interpersonal communication for information or advice in their decision-making about the adoption of a PV system. The study also showed that there is a positive relationship between WTP for a PV system and level of opinion leadership on PV adoption.

This positive relationship has policy implications in regard to the diffusion of PV systems. Governments seeking to promote the diffusion of residential PV systems most efficiently should consider policies that encourage opinion leaders to adopt early in the process of diffusion. Because potential adopters often ask opinion leaders for information or advice about PV systems in the course of decision-making about adoption, such policies are likely to spur diffusion.

How might a government make use of the results obtained above in designing such policies? Suppose that a government can make use of the two common policy instruments, investment subsidies and FITs, to incentivize households to adopt PV systems. The challenge facing the government is to set the levels of subsidy and FITs such that opinion leaders are incentivized to adopt PV systems preferentially.

Based on the results yielded in this study, the government should set a higher subsidy rather than a higher PV electricity price. To understand why, suppose that the government offers neither subsidies nor FITs. In this case, households with greater WTP for a PV system would be the ones that purchase and install them. If a subsidy were to become available, such households would purchase and install with greater financial ease than otherwise. As the subsidy becomes higher, a larger number of households with greater WTP would adopt PV systems. In contrast, suppose that instead of a subsidy, a PV electricity price is offered under FITs. The households that would be most likely to adopt in this case would not be the ones with great WTP but rather those expecting their systems to generate a large amount of PV electricity. As the PV electricity price becomes higher, a larger number of households expecting large PV power generation would adopt PV systems. In other words, a higher PV electricity price under FITs would distort the natural mechanism whereby the greater a consumer's WTP for a good, the more likely he or she is to purchase it. In view of the positive relationship between WTP and opinion leadership, this means that households with higher levels of opinion leadership would be more likely to adopt a PV system if larger subsidies are offered than if a higher PV electricity price is offered.

The significance of this study is twofold. First, the study extends standard diffusion theory, which centers on interpersonal communication, to the diffusion of a PV system. Second, it offers practical policy implications for the design of incentives to promote diffusion.

Further clarification of the role of opinion leadership in PV system adoption may depend on similar research being conducted at a future time when residential PV systems have been adopted more widely. PV system diffusion in Japan was thought to still be in an early phase when the survey was conducted, which suggests that a substantial proportion of all adopters had a higher level of opinion leadership than the general population. After further diffusion, the differences in levels of opinion leadership among adopters will be more distinct, enabling refinement of the results presented here.

In addition, alternative methods of identifying opinion leaders may merit further consideration, although such methods will present challenges. In the present study, opinion leaders were identified by two criteria: when a respondent had adopted, and whether he or she reported being asked for information or advice about PV systems. The use of this set of criteria is well-founded, as explained in Sect. 10.2.3. It remains to be seen, however, if the results are robust, that is, independent of the method used for identifying opinion leaders. However, as mentioned in Sect. 10.2.3, alternative methods for identifying opinion leaders, such as sociometric methods, informants' ratings, or observation, preclude the use of an internet-based questionnaire, making large-scale surveys impractical. One possibility for future study is to use an internet-based survey in combination with interviews to collect information directly from respondents.

As use of the internet in decision-making on technology adoption becomes more widespread, its role in the adoption of PV systems in particular becomes more important. Indeed, substantial use of the internet to gather information and even to ask for advice about adoption was observed in this study: 50.4% of survey respondents (246 respondents) reported using the internet to gather information in the course of deciding to adopt, and 14.8% (72 respondents) reported uploading information about their own PV systems. Further discussion of the role of the internet in decision-making with regard to adoption is beyond the scope of this study, but the topic may be an interesting one for future research.

References

Antonelli M, Desideri U (2014) The doping effect of Italian feed-in tariffs on the PV market. Energy Policy 67:583–594

Bollinger B, Gillingham K (2012) Peer effects in the diffusion of solar photovoltaic panels. Market Sci 31:900–912

Campoccia A, Dusonchet L, Telaretti E, Zizzo G (2014) An analysis of feed'in tariffs for solar PV in six representative countries of the European Union. Sol Energy 107:530–542

Cherrington R, Goodship V, Longfield A, Kirwan K (2013) The feed-in tariff in the UK: a case study focus on domestic photovoltaic systems. Renew Energy 50:421–426

David P (1966) The mechanization of reaping in the ante-bellum Midwest. In: Rosovsky H (ed) Industrialization in two systems. Harvard University Press, Cambridge, pp 3–39

Faiers A, Neame C (2006) Consumer attitudes towards domestic solar power systems. Energy Policy 34:1797–1806

Griliches Z (1957) Hybrid corn: an exploration in the economics of technical change. Econometrica 48:501–522

Guo X, Liu H, Mao X, Jin J, Chen D, Cheng S (2014) Willingness to pay for renewable electricity: a contingent valuation study in Beijing, China. Energy Policy 68:340–347

Jaffe AB, Newell RG, Stavins RN (2003) Technological change and the environment. In: Mäler KG, Vincent JR (eds) Handbook of environmental economics. Elsevier Science, Amsterdam p, pp 461–516

Jager W (2006) Stimulating the diffusion of photovoltaic systems: a behavioral perspective. Energy Policy 34:1935–1943

JPEA (2014) Statistics www.jpea.gr.jp/pdf/150219_deployment.pdf. Accessed 8 Apr 2015 (In Japanese)

Kramer D, Bigelow P, Vi P, Garritano E, Carlan N, Wells R (2009) Spreading good ideas: a case study of the adoption of an innovation in the construction sector. Appl Ergon 40:826–832

Labay DG, Kinnear TC (1981) Exploring the consumer decision process in the adoption of solar energy. J Consum Res 8:271–278

Lesser J, Su X (2008) Design of an economically efficient feed-in tariff structure for renewable energy development. Energy Policy 36:981–990

Martin N, Rice J (2013) The solar photovoltaic feed-in tariff scheme in New South Wales, Australia. Energy Policy 61:697–706

Mendonça M (2007) Feed-in tariffs: accelerating the development of renewable energy. Earthscan, London

METI (2013) A report on the diffusion of photovoltaic systems. http://www.meti.go.jp/meti_lib/report/2013fy/E002502.pdf. Accessed 30 Nov 2017 (In Japanese)

Moor GA (2014) Crossing the chasm: marketing and selling disruptive products to mainstream customers, 3rd edn. Harper Business, New York

Noll D, Dawes C, Rai V (2014) Solar community organizations and active peer effects in the adoption of residential PV. Energy Policy 67:330–343

Parker P (2008) Residential solar photovoltaic market stimulation: Japanese and Australian lessons for Canada. Renew Sust Energy Rev 12:1944–1958

Rai V, Robinson SA (2013) Effective information channels for reducing costs of environmentally-friendly technologies: evidence from residential PV markets. Environ Res Lett 8(014044):1–8

Rigter J, Vidican G (2010) Cost and optimal feed-in tariff for small scale photovoltaic systems in China. Energy Policy 38:6989–7000

Rogers EM (2003) Diffusion of innovations, 5th edn. Free Press, New York

Statistics Bureau of Japan (2010) Population census of Japan. www.stat.go.jp/data/kokusei/2010/index.htm. Accessed 21 Mar 2015 (In Japanese)

Schelly C (2014) Residential solar electricity adoption: what motivates, and what matters? A case study of early adopters. Energy Res Soc Sci 2:183–191

Shirai N, Masaoka K, Ohno K, Tokai A (2012) Analysis of residential photovoltaic generation, attention to characteristic of installed persons and factors of installation. J Jpn Soc Energy Resour 33:1–9 (In Japanese with English abstract)
Stern PC (1992) What psychology knows about energy conservation. Am Psychol 47:1224–1232
Sultan F, Winer RS (1993) Time preferences for products and attributes and the adoption of technology-driven consumer durable innovations. J Econ Psychol 14:587–613
Wilson C, Dowlatabadi H (2007) Models of decision making and residential energy use. Annu Rev Env Resour 32:169–203
Zhai P, Williams ED (2012) Analyzing consumer acceptance of photovoltaics (PV) using fuzzy logic model. Renew Energy 41:350–357
Zhang L, Wu Y (2012) Market segmentation and willingness to pay for green electricity among urban residents in China: the case of Jiangsu Province. Energy Policy 51:514–523
Zorić J, Hrovatin N (2012) Household willingness to pay for green electricity in Slovenia. Energy Policy 47:180–187

Chapter 11
Public-Private Partnership in a Biomass Project

Abstract Worldwide, an increasing number of projects have been implemented that use biomass discarded as waste—designated biomass waste in this chapter—as a renewable resource. Such projects, called biomass projects, involve various parties, including municipalities, private companies, consortia, and NGOs. The purpose of this chapter is to clarify the optimal organizational form of a biomass project. To begin, we survey examples of biomass projects in Japan to identify their typical features. Given that biomass projects have both public and commercial aspects, we are concerned with public-private partnership (PPP) as an organizational form for biomass projects. To examine the applicability of PPP to a biomass project, we review previous studies of PPP in the economics literature. The results are mixed: PPP is preferable under some conditions but not appropriate under others. These findings might be attributable to the method employed: the literature reviewed is only concerned with the contractual aspect of PPP, and our discussion is somewhat subjective. It may be useful for future research to take a different approach, recognizing that participants in a biomass project are all concerned with the affairs of the region and will thus collaborate on the project beyond making profits.

Keywords Biomass · Public-private partnership · Private finance initiative
Incomplete contract

11.1 Introduction

Biomass has been attracting considerable attention as a renewable energy source. One reason for this development may be its carbon neutrality: biomass does not increase net carbon dioxide if it is directly or indirectly used as fuel, and it thereby reduces greenhouse gas emissions. For example, ethanol fuel, which is produced from corn and sugar cane, and biodiesel, which is produced from vegetable oils and palm oil, are deemed carbon neutral. Alternatively, biomass may be burnt directly in a boiler to yield heat and electricity without an increase in the net amount of carbon dioxide. Furthermore, using biomass will contribute to launching new businesses in

Y. Yamamoto, *Feed-in Tariffs and the Economics of Renewable Energy*,
https://doi.org/10.1007/978-3-319-76864-9_11

a region, thereby creating job opportunities, as biomass is often used in the same local area where it is produced.

Biomass as a resource may be divided into two categories. In the first category, biomass is produced deliberately, mainly for use as input in biofuel production, as mentioned above. On the other hand, in the second category, biomass is yielded as by-products, residues, or waste; this category includes food waste, wood waste, sewage sludge, livestock excreta, agricultural waste such as rice straw, rice husks, and bagasse. This type of biomass may be transformed into heat, electricity, methane, biodiesel fuel (BDF), wood pellets, and so forth. In addition to products related to energy, compost may be produced. Accordingly, the use of biomass will contribute to alleviating both local and global environmental problems. In this chapter, we focus on biomass in the second category, hereafter designated as biomass waste or simply biomass. In the following discussion, the use of biomass is not limited to energy sources but includes compost production as well.

Although the use of biomass has many favorable features, some hurdles must be overcome when implementing biomass projects. For example, the equipment used in such projects generally remains costly despite technological developments, production costs are also high, mainly due to the high cost of gathering biomass, and the product is not economically competitive on the market.

To overcome these challenges, many central and local governments have supported biomass projects. For example, according to Kuzuhara (2005), the Japanese government has established an initiative called the Biomass Nippon Strategy, which aims to prevent global warming; create a recycling-oriented society; foster new strategic industries with a competitive edge; and support agriculture, forestry, and fishery, as well as associated rural communities. To implement the Biomass Nippon Strategy, the government established a system called the Biomass Town. In this system, a local community, designated as a biomass town, is supported in systematically, consistently, properly, and comprehensively using biomass to produce energy or materials by making better use of partnerships among relevant parties in the community. Financial support has enabled the development of biomass towns.

Although such support systems are crucial for biomass projects to be successful, building appropriate partnerships into such projects is also important. In fact, a variety of partnerships are formed by different parties, including—in many cases—municipalities, private companies, consortia, cooperatives, and NPOs. It is appropriate for both emitters of biomass waste and users of biomass products to participate in these projects. Furthermore, a municipality may choose to participate because utilizing biomass waste will contribute to the local environment and the economy.

Regarding partnership in these projects, public-private partnership (PPP) has been widely discussed (for example, Rosenau 2000; Bult-Spiering and Dewulf 2006). PPP is a long-term contract between a government and a private company to provide a public service. In PPP, all tasks in a project are bundled, including financing, construction, operation, and maintenance of the facility, and the public sector contracts with an entity in the private sector to undertake all of these tasks (Yescombe 2007; Engel et al. 2014). PPP has been used for many types of public service provision, including transportation, schools, hospitals, and prisons.

The question naturally arises whether PPP is applicable to biomass projects, and the purpose of this chapter is to address this question. We examine a biomass project that uses predominantly biomass waste and whose products include wood pellets and compost in addition to heat and electricity.

The remainder of this chapter is organized as follows. Typical features of biomass projects in Japan are surveyed in Sect. 11.2, which focuses on products, driving forces, and organizational form. In Sect. 11.3, PPP—as an organizational form—is discussed in a general context. Section 11.4 reviews the economics literature on PPP as a contract made between the public and private sectors. Following the literature review, in Sect. 11.5, we discuss the applicability of PPP to a biomass project. Section 11.6 summarizes the chapter.

11.2 Typical Features of Biomass Projects in Japan

In this section, before examining the applicability of PPP to biomass projects, we survey biomass projects in Japan to identify their typical features. Whether PPP is applicable will depend on those features.

To begin, it should be noted that there are at least two difficulties in surveying biomass projects. First, no database can cover all of the projects that use biomass waste. Many projects may be implemented on a small scale, and thus some are unlikely to be included in any database. Second, it can be difficult to determine whether a project is successful. For example, a project cannot be considered successful if it demands too much financial support from a government, thus placing a heavy burden on citizens. Although information on subsidies might be available, it is still difficult to judge the success of a project.

Accordingly, using websites, we survey biomass projects in Japan as widely as possible, without regard to their status as successful or unsuccessful. In addition to conducting a website-based survey, interviews are conducted, if necessary, with relevant figures involved in a project.

We focus on four types of biomass that are often utilized in Japan: food waste, wood waste, used cooking oil, and livestock excreta. The survey is conducted with a focus on three points: what is produced in a project, what are the driving forces that enable a project to be implemented, and what form of organization undertakes a project.

11.2.1 Products

In biomass projects, energy and materials are widely produced, including heat, electricity, compost, methane, wood pellets, BDF, and more. In particular, heat and electricity are often produced by incineration.

Compost, through composting, and methane, through methane fermentation, are produced from food waste and livestock excreta. Compost is more common than methane because composting is technologically and economically more feasible than methane fermentation.

The compost produced is, in most cases, sold and used in the same area where the biomass waste is discarded. Agricultural products produced with that compost are also consumed in the same area, which will then discard food waste. In other words, a material cycle is formed within a region—from biomass waste through agricultural production to food waste.

The methane produced is used to generate heat and electricity, which are mainly used in-house; if there is a surplus, it is sold to an electric utility at a favorable price under a feed-in tariff system. The methane is also used locally for fuel.

Wood waste discarded from sawmills is used to produce heat and electricity. In most cases, the heat produced is used in-house at the sawmills for drying wood. Wood pellets are also produced from wood waste. They are sold as fuel for pellet-burning heaters and boilers that produce the hot water distributed in the region.

BDF is produced from used cooking oil. Production is on a small scale because the production facility is usually small, collecting used cooking oil over a wide area is costly, and the BDF produced is consumed locally. Accordingly, there are a limited number of vehicles that use BDF.

11.2.2 Driving Forces

Three driving forces are identified: legislative enforcement, regional economic development, and business opportunities.

Legislation aiming to combat environmental problems has enforced the use of biomass waste. This enforcement may be attributable to the fact that the amount of waste has been increasing rapidly, thereby also increasing the cost of disposal, while the space available for dumping is becoming scarce. The Japanese government has established several laws related to waste management and recycling.

For example, under one law, the Waste Management and Public Cleansing Law, waste is categorized into industrial waste and non-industrial waste. Industrial waste is defined as specified types of waste emitted from manufacturing activities. Non-industrial waste is defined as waste other than industrial waste, which includes waste emitted from offices, restaurants, hotels, and similar establishments, as well as waste discarded by households. Industrial waste must be disposed of by those who emit it, whereas non-industrial waste must be disposed of by municipalities. Therefore, there are incentives for emitters and municipalities to reduce and recycle their waste. Similarly, another law, the Food Recycling Law, requires reductions in food waste in particular and promotes recycling as a form of composting and livestock feeding. Furthermore, a law regarding livestock excreta, the Law on Livestock Excreta Management and Recycling, regulates the management of livestock excreta, which, previously, had not been appropriately disposed of and was instead put directly onto agricultural fields or piled up out in the open, causing environmental problems.

Economic development is also a driving force behind the use of biomass waste. As mentioned in Sect. 11.1, biomass projects may contribute to the economic development of a region, and thus they are supported by municipal governments. Many rural areas are currently in need of economic development because their main industries have historically been agriculture, livestock breeding, or forestry, which are declining as the population shrinks due to increased aging. Accordingly, the annual budgets of municipalities have been shrinking due to diminishing tax revenue.

Additionally, some parties have found business opportunities in the use of biomass waste. For example, some private companies that want to pursue organic agriculture—which is highly valued among certain people—use compost made from livestock excreta for their agricultural production.

11.2.3 Organizational Forms

In many cases, municipalities are involved in biomass projects. Generally, biomass projects are strongly connected with particular regions, as biomass is often utilized locally, where it is produced. Accordingly, the municipal government plays a role to some extent.

There are several different ways in which a municipality can be involved in biomass projects. First, a municipality can directly undertake a biomass project. Second, an entity established by a municipality can undertake a project in the region. Sometimes, several municipalities establish an association to undertake a project in their area. Third, a municipality can contract a biomass project out to a private company, which the municipality then regulates.

In some cases, those who are involved in agriculture may participate in a biomass project. For example, farmers may jointly undertake such a project with the help of a municipality; or, an agricultural cooperative may participate in a project. Traditionally, an agricultural cooperative plays a role in agriculture-related activities, especially in an area where agriculture is a major industry. In other cases, an agricultural company undertakes a biomass project. These agricultural companies often pursue organic agriculture that does not involve the use of agricultural chemicals and chemical fertilizers.

Sawmills often undertake projects that use wood waste to produce heat, electricity, and wood pellets. In some cases, a group of sawmills, such as a forestry cooperative, builds a facility for such a project.

An NPO sometimes produces BDF from used cooking oil. This type of project is typically implemented on a small scale: small containers for stocking used cooking oil are distributed to households, restaurants, hotels, and similar establishments, and then collected at a certain interval.

11.2.4 Follow-up Discussion

Following the observations in Sects. 11.2.1 through 11.2.3, we identify three crucial features of successful biomass projects with regard to organizational forms.

First, the public sector should be involved in the biomass project, as the use of biomass waste will contribute to society—including the region where the waste is emitted—by alleviating environmental problems. Furthermore, the biomass project may facilitate local economic development. Accordingly, the public sector, particularly the municipality, has sufficient rationale for its involvement, direct or indirect, in a biomass project.

Second, a municipality can play a variety of roles in a biomass project. It may provide financial support, serve as coordinator or organizer, or directly undertake a project.

Lastly, the private sector should also be involved in the project. Biomass waste is often emitted as by-products, residues, or waste from business activities in the private sector. The economic efficiency of a project can be enhanced by involving emitters of biomass waste. Generally, a private company is likely to have substantial knowledge of both technology and management with regard to the use of biomass waste.

Therefore, a partnership between the public sector and the private sector will be important if a biomass project is to be successfully implemented, although biomass projects can take many organizational forms, as we saw in Sect. 11.2.3. The concept of public-private partnership or PPP has been widely discussed in economics and management science. In the next section, we will review PPP as an organizational form of public service provision in order to examine its applicability to biomass projects.

11.3 Public-Private Partnership

As discussed in Sect. 11.2.4, it may be preferable to have both the public and private sectors involved in a biomass project. In this section, let us consider the partnership between these two sectors in a general context.

The involvement of the public sector should be understood from the perspective of providing public goods and services. Public goods and services are defined in economics by two characteristics: non-excludability and non-rivalry. Non-excludability means that people cannot be excluded from consumption, and non-rivalry means that one person's consumption does not reduce the amount available to other people. It is usually argued that if public goods and services are provided by a private company, the level of provision will be lower than the socially optimal level due to non-excludability and non-rivalry. Accordingly, such goods and services should be provided by an entity in the public sector.

On the other hand, the involvement of the private sector should be understood from the perspective of business. A typical business structure consists of the procurement of inputs, production itself, and the sale of products. To make a profit in this process, efficiency is essential: appropriate production technology and managerial know-how should be employed. In this regard, the private sector will have advantages over the public sector.

In contrast, let us consider the possible disadvantages if public goods and services are provided by an entity in the public sector versus the private sector. First, suppose that an organization in the public sector undertakes such a project. The organization may put in less effort due to the problem of a soft budget constraint: a government will provide relief, such as a subsidy, to the organization if the project is not going well in order to avoid the failure of the project itself. Anticipating this, the organization may intentionally put in less effort (Qian and Roland 1998).

Next, suppose a government contracts a project out to a private company. In this case, the government cannot directly observe the actual state of the project in terms of whether the company is putting in its full effort. Accordingly, the private company might be tempted to manipulate the state of production to save effort. Anticipating this behavior, the government would order the company to produce at a lower level of output, leading to decreased profit for the company. In other words, the private company faces a trade-off between the low level of output ordered by the government and the cost of putting more effort into production; if the former effect is larger than the latter, the project will become less efficient (Schmidt 1996).

As an alternative way of providing public goods and services, PPP has been discussed in economics and management sciences. According to Yescombe (2007), PPP is defined as having the following four key elements. First, it is a long-term contract between a party in the public sector and a party in the private sector. Second, it includes design, construction, financing, and operation of an infrastructure or facility. Third, over the period of the contract, payments are made to the party in the private sector from the party in the public sector or from the user of the facility. Lastly, PPP specifies the ownership of the facility. In particular, the party in the private sector is paid based on the contract; as long as the service provision by the private party satisfies the level specified in the contract, the party is allowed to appropriate any surplus it produces in the project. Hence, the private party has an incentive to put substantial effort into the project.

There are many examples of PPP, especially in transport infrastructure, schools, hospitals, and prisons (for example, European Commission 2004). According to the website of the Public-Private Partnership in Infrastructure Resource Center (PPPIRC 2017), PPP is likely to be applied to investments in renewable energy technology. In fact, PPP seems to be a potential organizational form for biomass projects because it appears to offer both public and commercial features simultaneously.

Let us examine the applicability of PPP to a biomass project by reviewing prior works in the literature. While there are many studies of PPP itself in the economics literature, few studies examine PPP with regard to biomass projects. One such study is by Forsyth (2005), who conducted case studies of PPP in waste-to-energy projects in the Philippines and India. Forsyth argues that PPP works most effectively when investors, local governments, and citizen groups are willing to work together to implement a new technology and when they produce arenas that are locally inclusive in which to discuss the technology.

In the next section, we will review theoretical work on PPP, as there are few studies that address PPP in biomass projects.

11.4 Literature Review

The features of the use of biomass identified in Sect. 11.2 suggest that PPP could potentially play a role in biomass projects. In this section, we briefly review previous studies of PPP in the economics literature from a theoretical point of view.

11.4.1 Literature Review of Public-Private Partnership

To begin, we define PPP, which is also called a private finance initiative (PFI), before reviewing the literature. PPP is a method for delivering public services (De Bettignies and Ross 2004; Bennett and Iossa 2006). In a typical PPP project, the public sector agency, such as a local government, determines the required output specifications of a particular project. Then, this agency encourages bidding on the project, including its design, building, financing, and operation, to achieve the output specifications; it then makes a contract with the successful bidder, which may be a private sector organization or a consortium. The government agency then pays the successful bidder depending on the level of achievement. In addition, that bidder can make additional profit as long as it achieves the output specifications. Accordingly, the project should be efficiently implemented with the most suitable technology and management because the private sector organization chosen has an incentive to run the project as efficiently as possible to make a greater profit.

There are many studies of PPPs in the economics literature, including those of Grout (1997), Bentz et al. (2001), Hart (2003), Bennett and Iossa (2006), Martimort and Pouyet (2008), and Maskin and Tirole (2008). Among these, the studies by Hart (2003), Bennett and Iossa (2006), and Martimort and Pouyet (2008) particularly merit discussion as a starting point in understanding our problem.

11.4.2 A Study with a Model of Bundling Versus Unbundling

Hart (2003) investigates PPP in comparison with conventional provision of public services, developing a model involving an incomplete contract. In the model, a government aims to provide a public service, for which it makes a contract with a private party. The contract consists of two stages: a stage of facility construction and a stage of service provision. In this regard, the government may have two options: unbundling and bundling. In the unbundling option, the government makes contracts separately with two parties, one for the facility construction and another for the service provision. In this way, unbundling corresponds to conventional provision. On the other hand, in the bundling option, it makes a contract with a single party for the two tasks. Thus, bundling corresponds to PPP.

The contract specifies the basic standards of the facility in the case of unbundling (conventional provision) and those of the service in the case of bundling (PPP). The private party with which the government has made a contract is allowed to modify the attributes of the facility or the service by investing in an innovation without violating the contract. In this sense, the contract is considered incomplete.

There are two types of investment, i and e, in the facility construction stage, by which the running cost will be decreased. The difference is that investment i is a productive investment that makes the facility easier to run, whereas investment e is an unproductive investment that reduces quality. The party that constructs the facility incurs the total investment costs equal to $i + e$.

Then, Hart calculates the optimal investment levels in three cases: the social optimum, the optimum for unbundling, and the optimum for bundling.

It is supposed that the government seeks to keep the investment level as close as possible to the socially optimal level. The findings are that bundling is preferable for investment i and unbundling is preferable for investment e. Hart concluded as follows: PPP (bundling) is advantageous if—in the initial contract—the quality of the service can be effectively specified while the quality of the building cannot. In contrast, conventional provision (unbundling) is advantageous if—in the initial contract—the quality of the building can be effectively specified while the quality of the service cannot.

11.4.3 A Study with a Model That Includes Facility Ownership

Bennett and Iossa (2006) are also concerned with the level of investment that would be planned and implemented without violating the contract. Their model entails a government and two private firms, firms 1 and 2: the government supplies a public service by delegating to the two firms. Firm 1 specializes in facility construction and firm 2 in service provision. On the one hand, the two functions, facility construction and service provision, are said to be bundled if the two firms are organized as a consortium. On the other hand, the two functions are unbundled if the government contracts separately with each firm and the two firms operate independently.

Bennett and Iossa are concerned with the ownership of the facility in the model: if a firm plans to invest, it must negotiate with the owner about the investment. There are four possibilities with respect to who is the owner: firm 1, firm 2, the consortium, and the government. On the one hand, PPP corresponds to a situation where the government makes a contract with the consortium and the consortium has ownership. On the other hand, conventional provision corresponds to a situation where the government, retaining the ownership of the facility, makes contracts with firms 1 and 2 separately. It is assumed in the basic setting of the model that the owner of the facility will have full entitlement to the residual value of the facility when the contract expires.

The technicalities of the model can be described as follows. Notations here are defined anew regardless of the notations in Sect. 11.4.2. First, each firm may implement investment. On the one hand, firm 1 makes investment i, incurring cost i, at the beginning of the facility construction stage. Investment i improves the quality of the facility, resulting in increased benefit to society. However, the increased quality resulting from the investment leads to either reduction in cost, making service provision easier, or to augmentation in cost, requiring much maintenance. On the other hand, firm 2 makes investment e, incurring cost e, at the beginning of the service provision stage. Investment e increases the social benefit, with always-declining costs for service provision.

Bennett and Iossa calculated the optimal investment levels in terms of unbundling/bundling and ownership. Specifically, PPP corresponds to bundling, with ownership by the consortium, and conventional provision corresponds to unbundling, with governmental ownership.

The results by Bennett and Iossa (2006) are summarized as follows. On the one hand, when there is a positive externality of the investment in facility construction, that is, investment i reduces the running cost, bundling is always preferable. However, PPP (bundling with ownership by the consortium) is not always optimal; bundling with ownership by the government may be preferable under certain conditions. On the other hand, when a negative externality exists, in that investment i increases the running cost, and is not too strong, unbundling is preferred. Bennett and Iossa concluded, with other results, that PPP is likely to be preferred when the externality of the investment is positive, when the effects of the investment on the residual value of facilities are strong, and when the effects of the investment on social benefit are weak.

Similar results are obtained by Martimort and Pouyet (2008), who also investigated whether the building and management stages should be bundled together. They showed that when performance contracts can be written, if there is a positive externality between building infrastructures and managing the facility, both tasks should be performed by the same firm; however, with a negative externality, the two tasks should be split. It was also shown that if a contract is incomplete, PPP is preferable when there is a positive externality and the private benefits of owning assets are small enough.

11.5 Discussion

In this section, let us discuss the applicability of PPP to a biomass project following the results of the literature review in Sect. 11.4.

The study by Hart (2003) is concerned with the comparison between the bundling and unbundling of two tasks, construction and management. Bundling corresponds to PPP, whereas unbundling corresponds to conventional provision of a public service. According to Hart, the choice between the two options hinges on whether it is easier to write contracts on the quality of the building or of the service: if the quality of the service is easier to specify, PPP is preferable; if the quality of the building is easier to specify, conventional provision is preferable.

In a biomass project, the quality of the building is considered to be easy to specify at the time a contract is made because it depends, unlike the cases of prisons and hospitals, more on the choice of technology to be installed. On the other hand, the quality of the service will depend on the biomass waste supplied, which will be much more difficult to specify at the time the contract is made. Accordingly, based on the results of Hart (2003), conventional provision might be suitable for a biomass project.

The studies by Bennett and Iossa (2006) as well as Martimort and Pouyet (2008) are concerned mainly with the externality of the investment in construction. The findings common to these studies are that PPP is preferable with a positive externality, whereas conventional provision is preferable with a negative externality.

Let us examine whether the externality is positive or negative in a biomass project. Unlike schools and hospitals, what is important in a biomass project will be the quality of the biomass waste input. If a facility is equipped with suitable machinery when it is built, its products will meet a high standard, to some extent, even if the biomass waste supplied is of poor quality. In other words, the positive externality might exist in a biomass project. Hence, in this regard, PPP might be preferable.

Findings by Bennett and Iossa (2006) show that if the effect of the investment on the residual value of a facility is stronger, or if the effect of the investment on social benefit is weaker, PPP is more likely to be preferable. For a biomass project, these effects may be mixed because the effects on both residual value and social benefit might be large. Hence, in this regard, the superiority of PPP might not be clear.

In summary, according to the results of the theoretical investigation in the economics literature, PPP might not always be suitable for biomass projects. It should be noted that the discussion in this section is qualitative and subjective so that the results should be understood accordingly.

11.6 Conclusion

In this chapter, PPP was examined as a potential organizational form for biomass projects. We first surveyed cases of biomass projects in Japan to identify the typical features of such projects, and then we examined the applicability of PPP to biomass projects by reviewing theoretical studies of PPP in the economics literature.

It is found that although different parties undertake biomass projects, a municipality often plays a role in implementing these projects. Based on this observation, we argue that PPP is an organizational form that has the potential to implement biomass projects successfully. However, according to our review of the theoretical studies in the economics literature, the applicability of PPP to biomass projects is mixed: PPP is preferable under some conditions but is not appropriate under other conditions.

We should consider the results of the literature review while keeping in mind the following two points. First, the economics literature we surveyed focuses on the contractual aspect of the relationship between participants; in this relationship, there are two stages, building and management, and the dichotomy between PPP and conventional provision is mainly based on whether building and management are contracted with bundling or unbundling. However, from another point of view, the results might be different. In particular, if we focus on a situation in which a group consisting of a municipality, private companies, and local citizens is jointly seeking to reinvigorate a region, a different result may be obtained. This issue is discussed in depth below. Second, whereas the results of the theoretical investigation are quantitative and objective, the discussion about the implications of these results for a biomass project in Sect. 11.5 is qualitative and subjective. Hence, the results regarding applicability should be understood accordingly.

Nevertheless, we can draw some implications from the results. First, the survey results regarding examples of biomass projects in Japan can provide suggestions about the organizational form that will enable a biomass project to be successful. Most importantly, a municipality often plays a role in biomass projects because such projects are often implemented at a relatively small scale in a region, where the economy is not thriving and the population is small and aging. Then, the problem becomes how a municipality should be involved in such a project. As one possibility, we considered PPP, where a municipality and private companies jointly engage in a biomass project.

Second, as mentioned above, the applicability of PPP to a biomass project could not be assessed from a contractual viewpoint in the economics literature. Accordingly, another aspect should be investigated. As noted just above, biomass projects are often implemented in rural areas with small, aged populations, with the goal of regenerating the region. This may be one of the keys to a successful biomass project. Those who are connected with the region, such as the municipality, private companies in the region, and citizens, will be concerned with the affairs of the region. Accordingly, they may form a partnership that goes beyond making profits to achieve a common objective: revitalization of the region through a biomass project. If we focus on this aspect, another model, for example, public service motivation (Rainey and Steinbauer 1999; Francois 2000), should be developed to investigate the applicability of PPP.

Utilizing biomass waste will be increasingly important in terms of the mitigation of local and global environmental problems and the regeneration of regions. Partnerships among local governments, private companies, and citizens will play an important role in biomass projects, and further practical investigations will be needed.

References

Bennett J, Iossa E (2006) Building and managing facilities for public services. J Public Econ 90:2143–2160

Bentz A, Grout PA, Halonen M (2001) What should the state buy? CMPO Working Paper Series 01/40

Bult-Spieering M, Dewulf G (2006) Strategic issues in public-private partnerships: an international prespective. Blackwell Publishing, Oxford

De Bettignies J, Ross T (2004) The economics of public-private partnerships. Can Public Pol 30:135–154

Engel E, Fisher RD, Galetovic A (2014) The economics of public-private partnerships: a basic guide. Cambridge University Press, New York

European Commission (2004) Resource book on PPP case studies. http://ec.europa.eu/regional_policy/sources/docgener/guides/pppresourcebook.pdf. Accessed 30 Nov 2017

Forsyth T (2005) Building deliberative public–private partnerships for waste management in Asia. Geoforum 36:429–439

Francois P (2000) 'Public service motivation' as an argument for government provision. J Public Econ 78:275–299

Grout P (1997) The economics of the private finance initiative. Oxford Rev Econ Pol 13:53–66

Hart O (2003) Incomplete contracts and public ownership: remarks and an application to public-private partnerships. Econ J 113:69–76

Kuzuhara Y (2005) Biomass nippon strategy: why 'biomass nippon' now? Biomass Bioenerg 29:331–335

Martimort D, Pouyet J (2008) To build or not to build: normative and positive theories of public-private partnerships. Int J Ind Organ 26:393–411

Maskin E, Tirole J (2008) Public-private partnerships and government spending limits. Int J Ind Organ 26:412–420

PPPIRC (2017) The website of the public-private partnership in Infrastructure Resource Center. https://ppp.worldbank.org/public-private-partnership/. Accessed 30 Nov 2017

Qian Y, Roland G (1998) Federalism and the soft budget constraint. Am Econ Rev 88:1143–1162

Rainey HG, Steinbauer P (1999) Galloping elephants: developing elements of a theory of effective government organizations. J Publ Adm Res Theor 9:1–32

Rosenau PV (ed) (2000) Public-private policy partnerships. The MIT Press, Cambridge

Schmidt K (1996) The costs and benefits of privatization: an incomplete contract approach. J Law Econ Organ 12:1–24

Yescombe ER (2007) Public-private partnerships: principles of policy and finance. Butterworth-Heinemann, Burlington

Chapter 12
An Organizational Form for the Development of Renewable Energy

Abstract Many renewable energy (RE) cooperatives undertake local RE projects such as solar photovoltaic generation and wind-power generation, particularly in Europe. In contrast, in Japan, municipalities have recently become actively involved in setting up a type of company that will undertake such projects. This type of company will be called "a municipal RE company" in this chapter. The purpose of this chapter is to examine the effectiveness of this particular organizational form. A literature review, website surveys and an interview are conducted. The findings are as follows: first, the literature review on RE cooperatives reveals that one of the crucial factors for successful RE cooperatives is community involvement in the projects. Second, community involvement may be facilitated and achieved by public employees through public service motivation. Third, website surveys and an interview with a key figure at a municipal RE company confirm this view. It is suggested that a municipal RE company works, to some extent, in the same way as an RE cooperative and thus may be effective in undertaking local RE projects. Taking into account that the success of an RE cooperative will depend on historical, cultural, and legal conditions, a municipal RE company may be a potential alternative for a community that does not satisfy those conditions.

Keywords Renewable energy cooperative · Community involvement
Public service motivation

12.1 Introduction

Increasing attention has been paid to renewable energy (RE) development within regions, and particularly to generating and supplying electricity from RE sources

---

Modified, with permission of Springer Nature, from Yamamoto, Y. (2018), Optimal organizational forms for local renewable energy projects. In: Sayigh A (ed.) Transition Towards 100% Renewable Energy: selected papers from the World Renewable Energy Congress WREC 2017, Chap. 42, Springer International Publishing, Cham. I would like to thank Springer Nature.

Y. Yamamoto, *Feed-in Tariffs and the Economics of Renewable Energy*,
https://doi.org/10.1007/978-3-319-76864-9_12

within the region. This may contribute to the reinvigoration, revitalization, or regeneration of the region as well as to the alleviation of environmental and energy problems.

RE development projects are undertaken by various forms of organizations, including private companies, local governments, joint ventures, and cooperatives. One well-known form is that of community-based RE cooperatives (Schreuer and Weismeier-Sammer 2010; DGRV 2014; Tarhan 2015). This organizational form of RE projects is observed in Denmark, Germany, the Netherlands, and the UK, among others (Schreuer and Weismeier-Sammer 2010; Viardot 2013). In Denmark and Germany, RE cooperatives have been widely established (Huybrechts and Mertens 2014). According to DGRV (2014), the number of energy cooperatives within the DGRV (German Cooperative and Raiffeisen Confederation) has increased from 8 in 2006 to 718 in 2013. The prevalence of this organizational form has an historical foundation (Huybrechts and Mertens 2014). In other words, energy cooperatives might not be feasible in other countries that do not have the same historical background.

Recently in Japan, on the other hand, an increasing number of joint ventures, which consist of parties from the private and the public sectors, have been established to pursue local RE projects. The public sector, usually at the municipal level, plays a central role in these projects but does not undertake them directly as public works. Instead, it establishes a company that undertakes the project jointly with other private companies. The first case of this arrangement appeared in 2013, and to date, there are at least fourteen such projects in Japan. In this chapter, we call this type of joint venture "a municipal RE company".

The purpose of this chapter is to investigate the effectiveness of the municipal RE company as an organizational form for undertaking local RE projects. Recognizing that the RE cooperative, which is often observed in Denmark and Germany, is one of the successful organizational forms in local RE development, we investigate whether a municipal RE company will also be successful. We do this by first identifying a factor that is crucial to successful RE cooperatives and then examining whether that factor plays a role in municipal RE companies.

The remainder of this chapter is organized as follows. In Sect. 12.2, we look at several cases of municipal RE companies in Japan. Section 12.3 describes our methods for examining the effectiveness of a municipal RE company, including a literature review and follow-up surveys. The results of the literature review are presented in Sect. 12.4, and the follow-up survey results, consisting of website surveys and an interview, are presented in Sect. 12.5. Finally, Sect. 12.6 concludes the chapter.

12.2 Municipal RE Companies in Japan

As mentioned in Sect. 12.1, a municipal RE company has recently been established to undertake a local RE project in Japan. In this section, we look at several cases of these companies to understand what they are.

The first case is Company M in Municipality M, the population of which is approximately 14,000. This company was established in 2015 to pursue energy autonomy, that is, to help the municipality avoid depending heavily on electricity produced outside the region. Municipality M holds 55% of the shares in Company M. The company aims to produce and consume energy within the municipality, thus improving citizens' quality of life and reinvigorating the local economy. The company purchases surplus photovoltaic (PV) electricity from houses and also sells electricity to public facilities, independent businesses, and houses within the municipality.

The second case is Company N in Municipality N, the population of which is approximately 11,000. This company was established in 2016 to promote economic development in the region. Municipality N holds 41% of the shares in Company N. Company N purchases surplus PV electricity at a price higher than the national feed-in tariff and spends its profit on managing public infrastructure such as the water supply and on investments in projects that are aimed at revitalizing the region.

The third case is Company K in Municipality K, the population of which is approximately 23,000 people. It was established in 2016, aiming primarily at reinvigorating the local economy through self-sufficiency of both the energy supply and the economy. Municipality K holds 37% of the shares in Company K. Company K also purchases surplus PV electricity at a price higher than the national feed-in tariff and sells electricity to consumers with demands no lower than 50 kW.

What is common to these cases is that they are pursuing the reinvigoration and revitalization of the region rather than making a profit. It is also noteworthy that the municipalities have relatively small populations.

12.3 Methods

In Sect. 12.4, we first review the literature to examine the effectiveness of the municipal RE company as an organizational form. The literature review consists of two steps. In the first step, we look at RE cooperatives, which are often observed in Denmark and Germany and may be considered a successful organizational form. We seek to identify a crucial factor that makes RE cooperatives a successful organizational form for local RE projects.

Based on the results obtained in the first step, the literature review proceeds to examine whether this factor plays a role in the public sector; if this is the case, a municipal RE company may also be considered a successful organizational form for a local RE project because the public sector is involved to a greater extent.

After the literature review, we follow up on the results in Sect. 12.5. First, we survey the websites of municipal RE companies, and we then conduct an interview with a key figure at a municipal RE company.

12.4 Results

This section has two subsections. The crucial factor that makes an RE cooperative successful is identified in Sect. 12.4.1, where the discussion is extended into community involvement. Then, based on the factor identified in Sect. 12.4.1, public service motivation will be the focus of the literature review in Sect. 12.4.2. The interview with a manager of a municipal RE company—intended to explore the findings from Sect. 12.4.2—is described in Sect. 12.5.

Before presenting the results, let us clarify the definition and principles of a cooperative. The International Co-operative Alliance (ICA), an independent, non-governmental organization aiming to unite, represent, and serve cooperatives worldwide, cites seven principles of cooperatives: voluntary and open membership; democratic member control; member economic participation; autonomy and independence; education, training, and information; cooperation among cooperatives; and concern for community (ICA 2017). Among these, democratic member control and concern for community seem highly relevant to successful RE projects. These features may be summarized under the notion of community involvement. Thus, let us begin by examining whether community involvement plays a role in local RE projects.

12.4.1 Literature Review on Renewable Energy Cooperatives

Numerous studies address various characteristics of RE cooperatives (Schreuer and Weismeier-Sammer 2010; Viardot 2013; Yildiz 2014; Yildiz et al. 2015), and many studies emphasize the importance of democratic decision-making in cooperatives. Viardot (2013), for example, argued that cooperatives are democratically managed in that every member has an equal vote. Similarly, Yildiz et al. (2015) noted that the one-member-one-vote principle is one of the critical features of cooperatives. This decision-making system can encourage the members of a cooperative to play an active role in pursuing their objectives (Yildiz 2014).

The decision-making system also plays a role in tackling the so-called NIMBY (not in my backyard) problem. Schreuer and Weismeier-Sammer (2010) argued that local participation and ownership may reduce local opposition to the installation of wind farms; the installation of wind farms is often rejected by local citizens due to noise pollution, shadow casting, impact on the landscape, and other consequences (Musall and Kuik 2011). Viardot (2013) noted that the democratic decision-making process of a cooperative alleviates—in an equitable way—the concern over where a system will be installed.

There are also studies concerned with another aspect of cooperatives: community ownership of renewable energy facilities (Walker 2008; Warren and McFadyen 2010; Musall and Kuik 2011). According to Walker (2008), there is some evidence that if RE projects are owned or partly owned by the community, they will have fewer problems in obtaining permission from citizens and will be more locally acceptable. There are questionnaire-based surveys supporting the view that community ownership enhances public acceptance of local RE projects. For example, Warren and McFadyen (2010) showed that the wind farms owned by the community are associated with more positive local attitudes compared with farms owned by commercial companies. Similarly, Musall and Kuik (2011) showed that community co-ownership of wind energy leads to higher levels of acceptance compared with ownership by a commercial company.

Democratic decision-making and community ownership are the most important features of RE cooperatives, making RE cooperatives a successful organizational form for local RE projects. Further consideration allows us to summarize these two features under the idea of community involvement. In fact, Walker and Devine-Wright (2008) examined the policy and practice of community RE in the UK by constructing a database of projects connected to the community, interviewing policymakers and managers involved in community projects, and conducting case studies. Their findings were that more direct and substantial involvement of local people contributes to greater acceptance and support for a project.

The reason why community involvement is crucial to local RE projects may be that such projects appear to be deeply related to the identity of the community itself (Walker et al. 2010; Bomberg and McEwen 2012; Seyfang et al. 2013; Wirth 2014). Emphasizing the importance of a formalized cooperative system in the choice of organizational form, Wirth (2014) noted four community-related institutional forces that are central to the design of a biogas plant: community spirit, cooperative tradition, locality, and responsibility. Walker et al. (2010) compared two cases of local RE projects and concluded that trust between local people and groups that move projects forward can help projects work. Through case studies, Bomberg and McEwen (2012) found that a strong community identity and a search for local autonomy and community sustainability are relevant to community action concerning energy. Seyfang et al. (2013) conducted a web-based survey and concluded that the civil society basis of community energy groups and projects is fundamental to successful community projects.

To summarize the literature review regarding RE cooperatives, the most important factor in the success of a local RE project is community involvement in the project. Democratic decision-making and community ownership of RE facilities are considered crucial aspects of community involvement. In light of this understanding, let us examine whether community involvement is achieved in a municipal RE company in Japan.

12.4.2 Literature Review on Public Service Motivation

As we saw in Sects. 12.1 and 12.2, the most important characteristic of a municipal RE company is the active involvement of the public sector, i.e., a municipality. Hence, what we should examine next is whether the involvement of the public sector may play the role of community involvement; that is, whether a municipality serves as a proxy for its citizens.

To answer the above question, let us examine the motivation of public employees. It has been noted that public employees are motivated by the intrinsic value of their work (Francois 2000; Wright 2001; Jensen and Stonecash 2005; Besley and Ghatak 2005; Bright 2009; Huang and Feeney 2016). Jensen and Stonecash (2005) argued that one reason why workers in the public sector may derive utility from their tasks is that identifying with the mission of the public sector plays an important role in their work. Similarly, Besley and Ghatak (2005) stated that workers in a mission-oriented sector pursue the goals of the organization to which they belong because they perceive associated intrinsic benefits.

What, then, is the intrinsic value to be realized through work in the public sector? It may be strongly related to the notion of Public Service Motivation (PSM). PSM is defined as "a general altruistic motivation to serve the interests of a community of people, a state, a nation or humankind" (Rainey and Steinbauer 1999). Francois (2000) argued that there is considerable survey-based evidence that PSM exists. Bright (2009) found, through a questionnaire survey, that an important relationship exists between PSM and the intrinsic nonmonetary work preferences of public employees. Hence, it is more likely that PSM, which exists in the public sector, facilitates service provision for the public. Furthermore, Huang and Feeney (2016) found, using data from national surveys of local government managers in the US, that managers with higher PSM report greater levels of citizen participation in decision-making. Therefore, it will be PSM that enables the involvement of the public sector to realize the benefits of community involvement.

In summary, public employees are typically motivated by PSM to serve their region and citizens. Accordingly, we may consider a municipality to be a proxy for its citizens.

12.5 Follow-up Surveys

The results of the literature review are summarized as follows. Suppose that a RE cooperative is one of the successful organizational forms for a local RE project, on the grounds that many RE projects have been successfully implemented by organizations of this form, for example, in Spain and Germany. One of the crucial aspects of an RE cooperative is considered to be community involvement. The public sector may succeed at community involvement to a great degree because public sector employees are typically motivated by PSM. Therefore, a municipal RE company, in which a municipality is deeply involved, may succeed at community involvement to a great degree.

To supplement these results, we examine the extent to which municipal RE companies actually care about the region and its citizens.

As mentioned in Sect. 12.1, there are at least fourteen municipal RE companies. Surveying the websites of these companies reveals that every company aims to contribute to regional economic development, although the extent of their concern varies. Some companies aim to reinvigorate the region, and to this end, they seek to establish a system where electricity is produced and consumed within the region; others aim to establish such a system, hoping that it will contribute to the reinvigoration of the region.

A follow-up interview was conducted, revealing that a municipal RE company works for the region and its citizens like a municipality. A municipal RE company, anonymously called Company X in this chapter, was selected for the interview because the company is well known as a municipal RE company in Japan. A key figure of the company was interviewed in January 2017.

To begin, Company X is described in brief. It is located in Municipality X, which first established a foundation from which to undertake three solar photovoltaic power generation projects, promoting further RE development in the region. Municipality X owns 60% of the shares in the foundation, with the remaining 40% owned by a private company. The foundation established Company X in 2015 by investing 100% of the shares. In other words, Municipality X invests in Company X indirectly through the foundation. The manager of Company X had been a public employee at Municipality X, and the company's office is currently in the municipality building.

Two important results were obtained from the interview. First, Company X has been established to regenerate the region. This may be attributable to the current situation in Municipality X: more than 80% of the municipality is covered with forest. The population is nearly 17,000 people living in 6800 households and has been decreasing for at least 30 years. Those aged 65 and over constitute more than one-third of the population. Forestry and agriculture had been the main industries, but they are currently on the decline and no other industry has replaced them.

Second, Company X was established not to promote and achieve community involvement but rather to manage RE projects efficiently and promote further RE development. In other words, the establishment of the company, which was recommended by a consulting company, was done purely for managerial purposes. However, the company may secure the trust of the citizens because it is a not-for-profit company and has a strong connection with Municipality X.

To summarize the interview, Company X will demonstrate public service motivation by working on behalf of the municipality's citizens.

12.6 Conclusion

This chapter investigated why municipal RE companies involved in RE projects have attracted increasing attention in Japan, while RE cooperatives have successfully been undertaking RE projects in Europe, particularly in Spain and Germany. A literature review revealed that the important aspects of RE cooperatives are community involvement in RE projects and employees in the public sector who are generally motivated to work for the public; furthermore, municipal RE companies, which are strongly connected with the public sector, are likely to work for the public. These findings were confirmed by a survey of the websites of municipal RE companies and by an interview with a key employee of a municipal RE company.

The success of RE cooperatives is closely related to the fact that cooperatives are historically familiar in certain countries. On the other hand, there are many other countries, such as Japan, where the organizational form of the cooperative is less familiar. Furthermore, an RE cooperative in which many local residents participate is difficult to set up, partly due to legal conditions (Yamamoto 2016). In these countries, municipal RE companies may have the potential to undertake local RE projects successfully.

According to Tarhan (2015), local factors such as existing levels of trust, familiarity with cooperatives, and a history of cooperation affect the success of RE cooperatives. Similarly, the involvement of municipalities will be contextual. Accordingly, the organizational form that is best suited for RE projects in an economy depends on its economic, social, environmental, historical, and cultural characteristics (Tarhan 2015).

Interestingly, many cooperatives work with municipalities. According to DGRV (2014), for some 60% of energy cooperatives, a municipality participates in a cooperative as a member and actively engages in the cooperative committees. Schreuer and Weismeier-Sammer (2010) also noted that the relationship between a municipality, particularly the mayor, and active citizens is of crucial importance. Moreover, Seyfang et al. (2013) observed that community energy groups often work in partnership with other organizations, most prominently with local authorities. These observations suggest the importance of a municipality's role in RE development in a region, which may be related to the public-private partnership we discussed in Chap. 11. The role of municipalities in local RE projects is an important topic for future research.

References

Besley T, Ghatak M (2005) Competition and incentives with motivated agents. Am Econ Rev 95(3):616–636

Bomberg E, McEwen N (2012) Mobilizing community energy. Energy Policy 51:435–444

Bright L (2009) Why do public employees desire intrinsic nonmonetary opportunities? Public Pers Manag 38(3):15–37

DGRV (2014) Energy cooperatives: findings of survey conducted by the DGRV and its member associations, Spring 2014. DGRV (German Cooperative and Raiffeisen Confederation), Berlin

Francois P (2000) 'Public service motivation' as an argument for government provision. J Public Econ 78:275–299

Huang W-L, Feeney MK (2016) Citizen participation in local government decision making: the role of manager motivation. Rev Public Pers Adm 36(2):188–209

Huybrechts B, Mertens S (2014) The relevance of the cooperative model in the field of renewable energy. Ann Public Coop Econ 85(2):193–212

ICA (2017) What is a co-operative. https://ica.coop/en/what-co-operative. Accessed 30 Nov 2017

Jensen PH, Stonecash RE (2005) Incentives and the efficiency of public sector-outsourcing contracts. J Econ Surv 19(5):767–787

Musall FD, Kuik O (2011) Local acceptance of renewable energy: a case study from southeast Germany. Energy Policy 39:3252–3260

Rainey HG, Steinbauer P (1999) Galloping elephants: developing elements of a theory of effective government organizations. J Publ Adm Res Theor 9:1–32

Schreuer A, Weismeier-Sammerand D (2010) Energy cooperatives and local ownership in the field of renewable energy technologies: a literature review, RiCC-research report 2010/4, Vienna University of Economics and Businesses. https://www.wu.ac.at/fileadmin/wu/d/ri/ricc/Forschung/Laufende_Projekte/researchreport2010_4.pdf. Accessed 30 Nov 2017

Seyfang G, Park JJ, Smith A (2013) A thousand flowers blooming? An examination of community energy in UK. Energy Policy 61:977–989

Tarhan MD (2015) Renewable energy cooperatives: a review of demonstrated impacts and limitations. J Entrep Org Divers 4(1):104–120

Viardot E (2013) The role of cooperatives in overcoming the barriers to adoption of renewable energy. Energy Policy 63:756–764

Walker G (2008) What are the barriers and incentives for community-owned means of energy production and use? Energy Policy 36:4401–4405

Walker G, Devine-Wright P (2008) Community renewable energy: what should it mean? Energy Policy 36:497–500

Walker G, Devine-Wright P, Hunter S, High H, Evans B (2010) Trust and community: exploring the meanings, contexts and dynamics of community renewable energy. Energy Policy 38:2655–2663

Warren CR, McFadyen M (2010) Does community ownership affect public attitudes to wind energy? A case study from south-west Scotland. Land Use Policy 27:204–213

Wirth S (2014) Communities matter: institutional preconditions for community renewable energy. Energy policy 70:236–246

Wright BE (2001) Public-sector work motivation: a review of the current literature and a revised conceptual model. J Publ Adm Res Theor 11(4):559–586

Yamamoto Y (2016) The role of community energy in renewable energy use and development. Renew Energy Environ Sustain 1(18):1–4

Yildiz Ö (2014) Financing renewable energy infrastructures via financial citizen participation: the case of Germany. Renew Energy 68:677–685

Yildiz Ö, Rommel J, Debor S, Holstenkamp L, Mey F, Mükkerm JR, Radtke J, Rognli J (2015) Renewable energy cooperatives as gatekeepers or facilitators? Recent developments in Germany and a multidisciplinary research agenda. Energy Res Soc Sci 6:59–73

FSC
www.fsc.org
MIX
Papier | Fördert
gute Waldnutzung
FSC® C083411

Kolibri 360 GmbH
Ferdinand-Jühlke-Straße 7
99095 Erfurt, Deutschland
produktsicherheit@kolibri360.de